Resilience of Water Supply in Practice: Experiences from the Frontline

Resilience of Water Supply in Practice: Experiences from the Frontline

Edited by

Leslie Morris-Iveson and St John Day

Published by

IWA Publishing
Unit 104–105, Export Building
1 Clove Crescent
London E14 2BA, UK
Telephone: +44 (0)20 7654 5500
Fax: +44 (0)20 7654 5555
Email: publications@iwap.co.uk
Web: www.iwapublishing.com

First published 2021
© 2021 IWA Publishing

Disclaimer
The information provided and the opinions given in this publication are not necessarily those of IWA and should not be acted upon without independent consideration and professional advice. IWA and the Editors and Authors will not accept responsibility for any loss or damage suffered by any person acting or refraining from acting upon any material contained in this publication.

British Library Cataloguing in Publication Data
A CIP catalogue record for this book is available from the British Library

ISBN: 9781789061611 (paperback)
ISBN: 9781789061628 (eBook)
ISBN: 9781789061635 (ePub)

This eBook was made Open Access in December 2021.

Contents

Chapter 3
Transforming a water company to improve service levels and resilience: Lessons from Sierra Leone *39*
St John Day, Nitin Jain, Tom Menjor and Maada K Penge

Chapter 4
Mobilising the public to reduce household water use in Essex and Suffolk Water *59*
Fatima O. Ajia, Tim Wagstaff and Liz Sharp

Chapter 5

Water resources east: An integrated water resource management exemplar *81*

Nancy Smith, Robin Price and Steve Moncaster

Chapter 6

Implementing integrated water resources management locally in rural catchments: Lessons from eastern Sudan *99*

Khaled Mokhtar and St John Day

Chapter 7
Can and should refugees and communities that host them expect better performing and resilient water supply services? . . *121*

Ryan Schweitzer, St John Day, David Githiri Njoroge and Tim Forster

Chapter 8
Solar-powered water systems for vulnerable rural communities: Alleviating water scarcity in Iraq
Mohammed Al-khateeb and Ali Alkhateeb

Chapter 9
Economic resilience in water supply service in rural Tajikistan: A case study from Oxfam
Orkhan Aliyev

Editors

Leslie Morris-Iveson is a chartered environmentalist who has, since 1999, worked in both policy and practice in the water and environment sectors, with particular focus on risks and threats to water resources and water supply. Leslie has spent over a decade in field based and national roles in the water and disaster risk reduction (DRR) sectors in many countries across Asia, Africa and Latin America, with similar experiences in the UK and Canada. Through this, she has gained first-hand experience in responding to water threats in insecure environments and in promoting approaches that scale up access to water for marginalized populations at global and national levels. Leslie has worked with a wide range of international organizations and non-governmental organizations and is now a consultant, specializing in policy advisory support related to global challenges and risks that affect water availability and supply.

Dr. St John Day is a chartered water engineer and environmental manager who has worked extensively on infrastructure and institutional reform programmes throughout his career. His interests have always focused on water, especially in humanitarian, post emergency and development contexts. He has worked on major water supply, river engineering and flood alleviation projects in the UK and previously served on the Institution of Civil Engineers Advisory Panel for International Development. Over the past 22 years, St John has undertaken assignments in multiple countries in sub-Saharan Africa and Asia. He has worked with and advised national governments, multi-national agencies, international non-governmental organizations and the private sector, often with a focus on Fragile and Conflict Affected States. Prior to his engineering career, St John served in the UK's Parachute Regiment. He is now a Principal Consultant at IOD PARC.

Preface

This book aims to highlight the challenges in water supply that are faced in a range of locations, from industrialized, to fragile and conflict affected, and low- and middle-income countries. Water supplies globally are at risk from both slow and rapid onset threats. As a result, services need to become more resilient and those working on the frontline, more prepared to respond to threats rapidly. This book describes the different ways in which water suppliers respond to the challenges faced.

The book is the outcome of a motivation to bridge the gap between theory and practice: between the somewhat complex concept of resilience promoted in recent years, and how the concept is evolving on the ground. It is also the result of the belief that awareness or even learning can emerge by seeing how others (including those in vastly different contexts from your own) react and respond to challenges. It is hoped that the cases will stimulate ideas when you look at what professionals in a similar role to you are doing in a different part of the world, and by considering how they are stepping up to the challenge of integrating resilience into their operations.

By presenting the experiences of those directly involved in water provision, this book draws out some key challenges and the common factors involved in building resilience. Many practitioners are working under very difficult and testing conditions and are forced to operate some way below their desired performance levels. Meanwhile, other water suppliers that are performing well are under increasing pressure to maintain standards as uncertainty continues to increase. Therefore, the experiences presented have resulted in a wide range of resilience initiatives.

This book aims to contribute in a small way to the ongoing debate regarding how to make water services more resilient. It also aims to "give a voice" to the managers and practitioners who are working daily to deliver more durable services. We accept the scope, scale and innovation of what is taking place far exceeds this volume. It is hoped however, that some of the thinking and the cases that are presented – in plain language—will contribute to building a basic and realistic understanding of what different practitioners are doing to implement resilience into critical water services.

The book's introduction describes the challenge faced, who this book is for, and describes what resilience means for water supply. It brings out what resilient water services look like and outlines some key elements in building such services. In spite of the many resilient approaches in practice around the world, millions face the impacts when services are not resilient. Some of those impacts are described. The individual chapters of the book are then summarized to help highlight the key points to the reader. The chapters in this book describe differing contexts and are contributed by authors with extensive experience of the approaches described. Finally, the conclusions draw out some lessons from the cases and offer recommendations based on practice.

Acknowledgements

This book is the result of a collaboration between international water professionals, who have worked together from time to time over the last 15 years and followed similar career paths in conflict, natural disasters and other crises – both in the UK and in locations in Africa and South Asia. Having worked on water policy and on water supply and resource management projects on the ground where risk is a given, the book has been generated from a genuine interest to contribute to the emerging resilience agenda, in terms of water supply, and the desire to communicate how global concepts, or sector terminology are interpreted on the ground. The process began in late 2017, and the concept of the book has evolved over a few years, however, the focus has always been to document the experiences of frontline practitioners (and hence the use of the term in the title). It is our belief (and experience) that multiple threats to water supply are now an absolute given and our awareness of the situation is changing. We believe that "expert" views should come not only from people with a broad, global overview, but also from those with specific knowledge and experience in their particular location of action. Whatever can be done to de-mystify how we plan and act, and demonstrate what we've done (and reflect on actions) in a realistic way – not always the best and biggest and most well-known cases—should be considered essential sector learning.

As co-editors we extend our sincere thanks to those who supported this project, for their advice, assistance and encouragement. Specifically, the authors thank Mark Hammond (IWA Publishing) for the encouragement to develop the concept for this book. We are grateful to Mark and others at IWA Publishing for their support in making this book open-access, as it is our belief that field knowledge, no matter how imperfect, should be open. As such, we are pleased that our work on this has been not for profit. We thank them for their advice and patience.

Thank you to Alena Čierna (Ecorys) who has also supported this project with inputs and advice. We are also both proud associates of Richard Carter, and wish to thank

Richard for advice and patient support throughout, particularly on the review of the introduction.

Thanks to the peer reviewers of this book, and other professionals who gave guidance at various stages, many of whom have inspired the editors' work as well. They have included:

Dipankar Aich (Consultant)
Clarissa Brocklehurst (Consultant)
Vincent Casey (WaterAid)
Sue Cavill (Consultant)
Chris Cormency (UNICEF)
Richard Franceys (Consultant)
Stephanie Hurry (Waterwise)
Hussam Hussein (University of Oxford)
Emmett Kearney (UNHCR)
Harold Lockwood (Aguaconsult)
Neil Macleod (Consultant)
Gary Morris-Iveson (Consultant)
Phil Outram (United Kingdom Foreign Commonwealth Development Office)
Nathan Richardson (Waterwise)
Fiona Ward (UNICEF)

We have been inspired by a number of other water sector professionals, and are grateful for the many informal conversations we have had over the years, and the general collaboration in the water-risk sector.

Finally, the authors thank the contributors for their participation in this book, willingly sharing their experiences and their enthusiasm for this project, and for sharing knowledge, despite the turns the book has taken. We have had the privilege of working with many of the contributors to this book over several years. The book would not exist without the contributors.

Disclaimer

The content of this book reflects only the authors' views.

Chapter 1
Introduction

Leslie Morris-Iveson and St John Day

Uncertainty and interconnecting crises are no longer exceptional. In almost every part of the world, living with water crises is an everyday reality for many. Yet, water supply sustains a functioning society, and as such any threat to it must be countered head-on. Frontline water suppliers, routinely forced to respond and adapt so they can deliver water in the face of all challenges, have found that their task has become much more complex. They must accommodate growing populations that use ever-increasing amounts of water, and do so in the face of climate change, once a distant spectre but now an all-encompassing catastrophe demanding fast and adequate response. As the climate becomes increasingly erratic, water supply becomes increasingly unreliable and intermittent; couple that with long-standing performance issues caused by failing infrastructure or a lack of investment in maintaining and growing services, and it becomes obvious that utility management practices need to improve quickly. The problems of leaks and ageing infrastructure have become ever more demanding, and the coronavirus disease 2019 (COVID-19) pandemic has shown us that reliable and clean water provision is ever more critical to human life.

The front line of water supply has never presented more challenges than it does today. Resilience – the ability to anticipate variability, and deal with, recover from and learn from complex shocks and pressures – has quickly moved to the top of the agenda and has to apply to all water supply operations in all circumstances.

doi: 10.2166/9781789061628_0001

Solutions and opportunities to improve water security do exist. As well as technologies, there are decision processes, partnerships and other approaches which defy risks and ensure continuity of water flow during crises. The ability to share, interpret and use knowledge and experience supports the development of actions that respond to crises. To support those actively responding to realities on the ground, we need to reflect, learn and better understand how ideas of resilience can be translated into practice.

1.1 ABOUT THIS BOOK

This book is meant for anybody planning and delivering water supply services under increasingly difficult conditions. It spotlights resilience as a subject that needs focused attention because almost a quarter of the world's population now faces water crises (Hofste *et al.*, 2019). It explores actions and interventions on the ground that have advanced the provision of resilient water supply in response to complex challenges. The experiences of frontline water suppliers highlight the emergence of different practices, and the book explores a range of current methods that contribute to making water services more resilient.

From the moment we set out to create this book, we saw the vital importance of giving a voice to practitioners delivering water services on a day-to-day basis. Without knowledge of their local experiences, government institutions, the private sector and civil society cannot develop resilient service delivery models that can be supported and replicated elsewhere. Consequently, this book should interest everybody engaged in trying to strengthen the supply of water services in a range of locations and contexts. It will also be of interest to those undertaking professional qualifications because they will need to apply a strong thinker−practitioner approach.

The book draws on a range of locations and contexts to highlight the challenges being faced and adaptations being implemented; it covers industrialized nations with apparently high performance levels, robust institutions and utilities, as well as low- and middle-income countries, some of which are directly affected by armed conflict or decades of underinvestment. The experiences represented vary considerably, as do the available resources and institutional capacities.

In the process of editing this book, we have come to understand that much is being achieved in the struggle to provide safe water to entire populations in challenging circumstances, that building in resilience to climate change and other threats is evolving, and that much can be learned from sharing experiences. We have listened to others, who might be in vastly different situations, describe their choices and implementation of action. However, we do recognize the scale of the challenge. None of those to whom we have spoken would claim that their work has overcome the challenges faced; nor would any claim to have developed a fully resilient system.

The implications of demonstrating a range of contexts can be described in three different scenarios.

The first is seen in industrialized nations, whose water suppliers and institutions may already be operating relatively efficient water supply systems with high performance levels. The systems in which they operate will, therefore, be adjusting to some significant and specific risks. There might be an adequate level of resources, in terms of technical capacity and funding (including longer term financing to keep the options running), to ensure the options are sustainable. It is interesting to see how such initiatives might contribute single-handedly to these specific risks, and to understand the actual and potential direct effects of new practices and/or infrastructure.

The second scenario, seen in several countries, involves water supply delivered through largely inefficient systems, where practitioners are trying to improve performance levels at the same time as striving to strengthen the enabling conditions for more resilient water services. These locations might demonstrate how practitioners deploy and balance different approaches, according to needs on the ground, while at the same time improving water services that adequately meet the needs of populations today and into the future. In this scenario, the process of building systems includes developing system-level capacity (human, financial, technical).

The last scenario involves countries in which water supply services are partially or entirely lacking. In these locations, building resilient water supplies is complex, challenging work that will take many years, even decades, to achieve. Actions taken are often primarily community based, and capacity building of national institutions and securing greater human, equipment and financial resources are a continuous struggle.

Practitioners and decision-makers must focus on immediate actions that will achieve the basics of adequate water provision, but how are they to make a water supply system more resilient when the network of people, institutions, infrastructure and resources necessary to deliver sustainable services does not yet exist? This book aims to share insights into frontline practitioners' actions and initiatives when they are pressured to improve water supply services and keep them working.

Some of the cases represent major changes in methods of water supply, such as the transition from fossil fuel to solar as an energy source in rural Iraq, although the scale of their achievements is currently limited. Some focus on strengthening high-quality service delivery work; examples include sound analysis, planning, design and construction, all coupled with high-quality supervision, as in Tajikistan. Others focus on more specific, though still important, aspects of making water services fit for the future; one such example is the reduction of water demand in Suffolk to address impending water scarcity.

All the cases are linked with a common thread: water suppliers are trialling and implementing innovative practices with the aim of attaining two objectives – to supply adequate water to growing populations and to make the supply system resilient to current and future challenges. These approaches will continue to adjust and adapt, guided by changes on the ground and by actors' and institutions' growing ability to develop capacities to absorb shocks so they can return systems to a pre-crisis situation. A summary of each chapter's content is provided at the end of this introduction.

Contributors to this book are all *practitioners*, professionals working for cities and local authorities, water utilities, water catchment (or basin level) management institutions and other government institutions, as well as for non-governmental organizations (NGOs), wider water service support institutions and partners. We hope that, by bringing all these practitioners' experiences and insights together in this book, we will not only help contribute to a realistic understanding of what 'working towards resilience' looks like in practice, but also stimulate informed interactions between service providers and service authorities. The book does not aim to provide an exhaustive rendering of the entire picture – the scope, scale and innovation of events far exceeds its capacity to do so – but provides a glimpse of useful cases. Above all, the book aims to contribute to practitioners' already vibrant learning and collaboration and to promote further understanding of the need to build resilience into critical water services.

1.2 WATER SUPPLY RESILIENCE: CONCEPTS AND CONSIDERATIONS

This section defines some of the terms and concepts used in the context of this book.

Resilience, the main theme of this book, can be described as the ability of the water service as a whole to anticipate variability, absorb and recover from different shocks as they happen and return to a pre-crisis or even achieve an improved position, and learn from the experiences.

The threat of water supply shocks is significant, as unmanaged shocks seriously disrupt water services, shutting down supply for hours, weeks or even permanently. Where excess risks are experienced and where water services are insufficient, a 'tipping point from high to low services and collapse' (Krueger *et al.*, 2019) can occur. The *tipping point* (or *threshold)* refers to the point at which disruption to the status quo and opportunities to transform the water service become much more difficult.

Institutions and systems with *adaptive capacity* are more able to accept the inevitability of unpredictability. They know that shocks and changes will occur and will be better equipped to plan, deliver and adapt water services, constantly inter-linking and innovating to accommodate the changing set of conditions.

Concepts relating to threats to and resilience of water supply are most vividly illustrated in the recent Cape Town Day Zero water crisis 2018, described in Chapter 2, where the impacts of severe drought conditions over three years led to major water shortages. As late as 2014, the six dams feeding the city were full. However, the effects of severe drought led water levels to decline drastically, eventually falling below 13.5% capacity (City of Cape Town, 2018). In early

2018, it was predicted that if no changes were made (or no additional rainfall recharged the dams), Cape Town would experience 'Day Zero', a day when the city would run out of water. This led to the initiation of severe water demand restrictions. Since the disaster of running out of water was avoided, the city has put a municipal-approved strategy into place to manage and diversify bulk water and reconcile supply and demand, as a way to build resilience into water supply. The multitude of actions described are all measures to adapt to future unpredictability.

Water service resilience can possibly be compared (or confused?) with sustainability. As WaterAid has described (2011) sustainability is defined by Len Abrams *et al.* in *Sustainability Management Guidelines* (2000) as: '*Sustainability is about whether or not WASH services [...] continue to work over time. No time limit is set on those continued services [...]*'. Note that 'sustainable' describes something 'long-lasting'. Resilience describes something that has the characteristics that will allow it to overcome challenges and consequently last well into the future. We acknowledge that the term resilience is sometimes contested. There can be doubt about terminology such as *climate resilience* and different viewpoints as to how the term is used alongside the concept of sustainability. To what extent do development interventions really tackle current and future challenges? And what has been learned about planning to make infrastructure and services more resilient? These are questions at the centre of this book.

The focus of this book is on water services: the provision of water by suppliers, alongside the policies and institutional arrangements that enable water to flow, to households, businesses, industries and other user groups.

While sanitation is not covered widely in this book, we recognize that systematic resilience and sanitation is of vital importance for populations, alongside water supply, especially where needs are not met. We also recognize that sanitation provision and wastewater treatment are intimately connected to water supply, in that they are also provided by the same service providers. When sanitation is inadequate at a household or community level, this is a risk in itself. Where wastewater is untreated, particularly when in close proximity to water supply, it can contaminate valuable reserves of water. The term sanitation is also often associated with the process of treating and disposing of wastewater. As water companies become more proficient at supplying larger quantities of water, increased volumes of wastewater will become an issue of growing importance. The mandates of water utilities will need to evolve to prevent wider public health crises.

1.3 WHY RESILIENT WATER SUPPLY NEEDS TO BE PART OF THE NEW NORMAL

Throughout the world, pressures (situations which develop over time) or shocks (sudden-onset events), all have the potential to disrupt water supplies and exhaust institutional resources. Natural disasters, water scarcity, and deteriorating infrastructure represent some of the many crises being faced. These crises are complex and multi-dimensional on an individual level. However, in an ever-more connected world, they have the potential to interconnect and cascade. These changes are exacerbated significantly by the climate crisis in which immediate impacts are worsening by the day, presenting unprecedented challenges for the water supply sector. The scale and complexity of these challenges demand that service providers urgently consider the resilience of their water supply systems: how services are and will be impacted by these changes; and how customers' actions and behaviours must also be part of the changes that are necessary.

The risk of water and sanitation service failure due to climate change is increasing (Howard *et al.*, 2016), with water services gradually declining as a result of not being able to bounce back following successive shocks. Entrenched problems are now inevitable, whether sudden or continuous, and will have the effect of undermining sustainability and resilience. Climate change exacerbates water availability, and variability will only increase and become more frequent and severe. Droughts and floods are examples of increased water variability that present shocks to water systems. Drought hugely impacts on water services, not only lessening availability for supply, but degrading water quality due to diminished flows and reducing the flushing rate of water bodies. Flooding can physically destroy water infrastructure, but can also contribute to saline intrusion and water pollution where inundation of water sources occurs, flooding of infrastructure including latrines, and poor water quality due to sedimentation. Rising temperatures are making dry regions drier, and lead to less water being available for water supply, as high levels of evapotranspiration increase the risk of drought or prolong periods of drought. When water use increases in drought-prone areas, groundwater and surface waters are depleted. Extreme temperatures can also contribute to water quality problems, for instance, increasing algal blooms.

Climate change is a very clear driver of change to water supply provision, but population growth, increased water demands and land degradation are all concurrent and overlapping trends alongside climate change. Climate change should be viewed alongside other pressures and factors that currently undermine sustainable service delivery. In many countries, serious disruptions to infrastructure and service delivery already occur, regardless of the impacts of climate change. Immediate actions and measures are required to improve performance levels and are fundamental if service providers are to respond to climate change pressures.

Water scarcity in particular has become an unprecedented global threat to water provision, requiring immediate and long-term sustainable solutions. As populations increase, causing an increase in the demand for water, water scarcity can become endemic during the dry season, or periodically throughout the year. In conditions of water scarcity the resilience solution has often been to focus on harnessing available water resources through the construction of reservoirs, dams and more efficient supply systems. However, physical infrastructure must also be accompanied by sound stewardship of water resources and management arrangements need to be continuously reviewed.

In the UK, known for its high average rainfall, the threat of water scarcity led Sir James Bevan, Chief Executive of the Environment Agency, to warn of the 'jaws of death' in a speech in 2019, to describe a situation where, in the near future, there will not be enough water to meet needs (Bevan, 2019). The speech warned of a situation where water demand increases as the population, homes and business increase, while available water to supply those demands decreases, compounded by the effect of climate change. In the UK contexts, a range of different approaches are being implemented in the drier regions as described in later chapters, including managing water demand in Suffolk and Essex and working collaboratively in partnership across sectors to safeguard a sustainable supply of water, including on specific watershed restoration projects as described by Water Resources East in East Anglia.

Challenges faced by, but not limited to, low- and middle-income countries when attempting to make water supply more resilient are considerable. Water supply infrastructure may be old and crumbling or absent altogether. Major water supply systems may have suffered from a chronic lack of strategic planning and the finance needed to make improvements over many decades. Water companies and institutions may not have the resources and may struggle to perform essential maintenance functions, resulting in reduced supply efficiencies and low levels of customer satisfaction. These companies often also struggle to generate revenue and attract investment finance, and receive limited government support. In these cases, interventions must focus on being both appropriate and effective so they contribute to building resilient services in an incremental manner. If the right interventions are not identified or delivered professionally then they will not stimulate the necessary change.

In the situations described above, capital expenditures and ongoing costs required to keep water flowing are constantly increasing to meet growing needs. Unplanned urban sprawl, for example, is a particular challenge. Migration from rural to urban locations is a major trend throughout the world (United Nations Economic & Social Council, 2018) as people search for jobs and improved income security, leading to the emergence of informal settlements. Freetown in Sierra Leone has more than 70 unplanned settlements. Water companies like Guma Valley Water Company may be mandated to serve these populations, but they often lack the resources, infrastructure and finance to extend services and protect critical water and land resources. In addition, it is often the case that those wanting to improve their income security by moving to cities have little or no means with which to pay for water, resulting in water providers being unable to collect full payment for supplying water to these communities.

Against these challenges, water companies and mandated authorities will need to respond by building resilience into their services and infrastructure. In order to do this, customers, service providers and governments will need to work together to rapidly respond to existing and future challenges. Any planned changes will need to be relevant and implemented effectively.

1.4 IMPLICATIONS FOR PEOPLE WHEN SERVICES ARE NOT RESILIENT

Everyone, from individuals to households, needs water that is reliable, clean and safe to use, to support basic needs and public health. People also want to be assured that they will have reliable water supply today and for future generations, without harming the ecosystems which they rely on. As recently brought into clear focus during the response to COVID-19, many water utility companies set about connecting customers previously not able to directly access clean water due to their limited ability to pay, in recognition of the protection a reliable water supply provides in the fight to prevent the further spread of the pandemic. In the UK, as a direct result of the role water utilities played in quick and effective service provision during the pandemic (ongoing at the time of writing), the UK Government, including the Water Services Regulation Authority (Ofwat), requested that service providers expand their ambition and play their part in green economic recovery efforts, for a more resilient future (DEFRA, 2020).

When services are not resilient, the resulting impacts can be either short or long term and greatly affect how populations access water. Water services can often emerge as intermittent supply (i.e. water that is supplied for less than 24 hours a day, or water that is supplied at irregular periods of time during crises). Water trucking – or water delivery through trucks – can be the result when solutions fail, during drought or during crises. This is an example of a band-aid solution, normally used for emergencies, that has now become the new normal in some parts of the world. Supplying water intermittently or through modes which are unsustainable and costly (such as water trucking) are undesirable and provide a means of service which is unsatisfactory and unrelatable for those who receive it. In many cases, water is not provided at all. When services are insufficient, a variety of coping mechanisms emerge at the household. This can include storing and hoarding water, bartering and bargaining for water at different points of the day when there are needs. Those who have the means buy booster pumps, or drill private boreholes. Women are often affected the most in low-income settings, being responsible for caring for household members, and the elderly and sick and their hygiene needs.

Water crises, in varying forms, often disproportionately affect the most vulnerable in society. Delivery of water and sanitation services themselves confer resilience on people by providing the health and wellbeing benefits of a clean supply of water, safe removal of insanitary waste and the associated hygiene benefits. The provision of clean water and sanitation services supports coping capacities during the most critical times, such as when populations are dealing with disasters. Again, in low-income countries especially, lack of access to a clean water supply at or near their home particularly affects women and children, who are often the collectors of water when it is not supplied directly to their homes. This lack of access to a clean water supply has impacts on health, education and, ultimately, life opportunities.

Access to clean water and adequate sanitation is a universal human right, recognized by the United Nations General Assembly, through Resolution 64/292 (Resolution adopted by the United Nations General Assembly, 'The Human Right to Water and Sanitation', A/RES/64/292, 28 July 2010, available at: https://undocs.org/en/A/RES/64/292), which applies during crises, but also during all other non-crisis periods. The Resolution states that water is to be provided in a non-discriminatory manner and determines that states bear the primary responsibility to respect the right to adequate water and sanitation provision for all individuals within their territory. These rights have been further enhanced through the determination that *'the water supply for each person must be sufficient and continuous and safe and acceptable, physically accessible and within safe reach for all sections of the population, taking into account the needs of particular groups, including persons with disabilities, women, children and the elderly.....'*. Water services must also be affordable, allowing access to all, whatever their means.

1.5 IMAGINING A RESILIENT WATER SUPPLY

Taking into consideration the diverse challenges and perspectives mentioned above, we ask the question: what would a more resilient water supply service look like? Intuitively, it would be less prone to recurrent breakdowns and failures. It would also be more resistant to external shocks, such as drought or flooding. It would be robust, but also able to adapt to changing conditions that is to say water would continue to be provided even during drought periods, as an example. This implies not only a focus on robust infrastructure but also a diversity in schemes and water sources to support different performance pathways to spread risk and buffer shocks. Infrastructure that augments supply, or 'hard' interventions must be complemented with soft path actions – the many decisions and actions focused on productivity and needs of end users, that all build resilience and sustainability. Actions which foster collaboration in the water catchment, and raising awareness of stakeholders on actions that build resilience is needed. The example of Water Resources East, illustrated in Chapter 4 is an exemplary model of how this can happen in practice – bringing together diverse stakeholders in a water-stressed catchment to not only plan, but also act together.

There is also a consideration of focus and scale; which part of the supply system could be improved to provide the largest positive effect on building resilience into the supplies. This is subject to debate, but there is a stronger understanding of the distinct shocks and pressures facing water supply at the local and regional levels.

Often, smaller and more decentralized actions are the most appropriate as they can be managed locally by the people and institutions most affected by current and emerging shocks and stresses to their water supplies. The example illustrated by Oxfam in Tajikistan as one of the actors in decentralizing water supply in response to a highly centralized and inefficient system, shows us the major challenges in achieving this where governance is weak. This example is illustrated in Chapter 6.

The readers of this book will have their own experiences and views on resilience. We believe a resilient water supply encapsulates all the essential components that are required to maintain water delivery at the desired performance levels. These components include people, their norms and practices, policies, institutions, infrastructure, environmental resources and finance. It is unlikely to be achieved through a single intervention, instead it requires a way of working that recognizes that the working environment is often complex and full of uncertainty.

Concepts such as 'climate proofing', 'adaptive management' and 'build back better', and tools to guide practitioners in understanding water risks and making decisions during times of uncertainty, are all examples of resilience planning in action. (A large number of examples exist, such as: Deltares' Dynamic Adaptive Policy Pathways, see: https://www.deltares.nl/en/adaptive-pathways/; WWF's Water Risk Filter: https://waterriskfilter.panda.org; and WRI's Aqueduct Water Risk Atlas are a selection of useful examples that provide data or guidance.) Approaches such as vulnerability and risk assessment in the water sector and the water safety planning approach (WHO's water safety planning approach: https://www.who.int/water_sanitation_health/water-quality/safety-planning/en/), identify and prioritize risks to water supply, allowing targeted actions to be taken by managers. In their own way, such guiding concepts, tools and approaches all aim to protect populations from the future effects of climate change, social and economic adversities and environmental degradation. These are all seen as important approaches to make water resources, physical infrastructure, institutions and finance more sustainable and durable.

Another aspect concerns the *relevance* of interventions. If operators are not addressing the fundamental issues at hand, their approach is likely to be flawed. If water demand exceeds supply for example, and no new water resources have been harnessed and developed, or consumer water conservation and efficiency of use addressed, a water company may never be able to satisfy their customers' requirements. Consequently, interventions must address critical operational challenges. If the 'right' interventions are not implemented to high professional standards this could also have a profound adverse effect on sustainability and resilience. This is a pressing issue in many low-income countries, where the challenges of non-functionality and rapid decline of infrastructure services are well documented. In the future, it is feasible that governments and societies will demand assurances that service providers have the requisite skills to build resilient water services. They will need to be well designed, built and maintained by people that have the requisite skills, knowledge and experience to do so. This growing demand has been identified by the United Kingdom's Institution of Civil Engineers (ICE, 2018). This means standards of professionalism in planning, designing and implementing water provision services will need to improve.

It is worth highlighting that water supply infrastructure is not maintenance free. All water infrastructure is prone to breakdown or service disruption at some stage. In such difficult circumstances some overarching factors are particularly important. Wherever possible, measures need to be put in place to deliver high-quality, professional services from the outset. If services are poorly constructed then there will be little prospect of them remaining resilient. When breakdowns or interruptions occur, systems need to be put in place to ensure a rapid response. There also needs to be continuous improvements to the wider 'system' – the network of policy, legislation, people, institutions, infrastructure, finance and resources. These improvements need to be guided by a clear investment plan that serves as a directional compass.

Engaging with complexity requires the ability to visualize the issues at hand and break them down into their component parts so that any action focuses on areas of greatest positive impact. Focusing on technology alone to address discrete risks will only solve one side of the problem. Resilience approaches require different processes, including integrated approaches that reflect on the interdependencies between the different risks experienced or predicted.

1.6 FROM THEORY TO PRACTICE

Resilience on a wide scale can be achieved. One example is taken from The World Bank's Resilient Water Supply and Sanitation Services Report (2018) report, which provides a useful overview of Japan's response to the 2011 earthquake and associated tsunami. The report shows how Japan has used an operational framework to strengthen the resilience of water supply which was effective during the crisis. The framework highlights a number of important factors, including demonstrating that countries need strong legal and institutional frameworks. These are required so that high quality of professional design, construction and supervision standards are followed and contingency funds are set aside for emergency response. Roles and responsibilities for disaster risk management (DRM) must also be set out in detail. In Japan, water supply and sanitation systems planning have formed part of a wider Masterplan, with redundancy built into the system so that water supplies can be re-established rapidly after disasters. There has also been strong emphasis on engineering design, materials, construction quality and routine asset management planning. This should ensure there is routine asset management and vulnerable assets are identified and replaced routinely as a result of adequate capital maintenance expenditure. This plan also places a high level of emphasis on contingency planning, including developing funds and subsidy programmes for assets which build system resilience. These resources are set aside to improve systems and infrastructure, making them more disaster proof, so that the potential adverse impacts from a major disaster or crisis on the water supply are reduced. This helps to ensure there can be a timely and effective response and minimal system downtime.

The concept of 'Build Back Better' is one example of a resilience strategy aimed at reducing people's vulnerability to natural disasters – post emergency. The strategy forms part of Japan's DRM approach, and it emerged from the United Nation's Sendai Framework for Disaster Risk Reduction in 2015. This includes measures to improve physical infrastructure, social systems, economies and the environment. The idea of building back better is discussed in Chapter 7, in possibly the least resilient of contexts in the world. The chapter looks at making water supply resilient in long-term refugee camps that have been entirely reliant on external aid.

1.7 RESILIENT WATER SUPPLY

A key transition from simply building services to ensuring resilient services is to introduce a conceptual framework and a set of actions to ensure that the required parts of the process are being met. Figure 1.1 broadly illustrates six key areas and some example actions that promote resilient water supplies. The actions were developed from a brief literature review (GIZ, 2020; UNICEF, 2021) and from our own experiences of the different elements that need to be considered to deliver resilient water supply services. Many of the actions identified focus on environmental sustainability and sustainable water use and are therefore aimed at local, regional or national level as relevant.

Firstly, different actors must collaborate and plan together. Water supply must not be planned in isolation by the engineers who build the services. The critical threats to water supply should be evident, and actors should be able to correctly sequence activities to address risks. This recognizes the need for integrated approaches and reflects the interdependencies between water supply and other sectors that depend on large water quantities. By working with other water user groups, including in the agriculture and industry sectors, to jointly model and plan for future water resource needs, including scenario planning, the impacts of climate change and other pressures on shared water resources can be better mitigated. Crises and disasters happen in many forms and all infrastructure systems will fail at some point. When this happens there needs to be a process of analysis and recovery that can respond rapidly. The response must also ensure support for externalities when the magnitude and exact nature of future shocks cannot be predicted, for instance in Japan, entities responsible for coordination and enforcement of provisions are made clear through legal structures (The World Bank, 2018). This has enabled quick mobilization of external assistance.

Recognizing the importance of the water resources which are critical to meet water supply demands, actions are needed at the catchment level to ensure

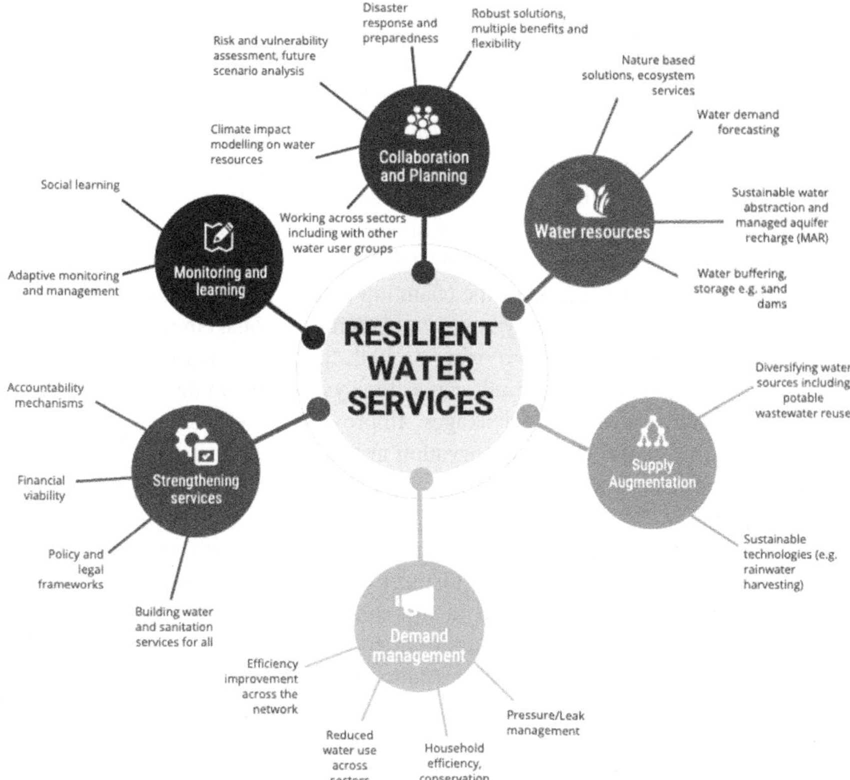

Figure 1.1 Conceptual framework for building resilient water services, with example actions under the six headings. *Source*: Leslie Morris-Iveson and St John Day.

reserves of water are managed sustainably. This includes, for instance, nature-based solutions that manage and restore natural ecosystems that enhance water availability and improve water quality. It is also necessary to understand renewable water availability and demand forecasting, estimating volumes of water that are required to meet future water demand. To provide a buffer against the impact of droughts, implementing sustainable storage methods at the appropriate scale (such as the use of sand dams in rural areas, typically in lower income countries) for water supply during dry periods. Sustainable methods for recharging aquifers (or 'managed aquifer recharge' methods) can be achieved by, for example, directing flood waters to naturally recharge aquifers which are being over-used, or to introduce treated wastewater under closely managed conditions and in specific circumstances.

Delivering macro- and micro-scale physical water supply infrastructure based on sound planning, design, construction and supervision is necessary to augment water supplies in response to population growth. Planning is critical as it plays a significant role in determining the resilience of water services post-construction. Services need to be constructed to high professional standards, based on designs relevant for the location and environment in which they will operate, with critical infrastructure and equipment being renewed routinely. Schemes to augment water supply should be robust but should be planned with a recognition of supply risk, including the specific impacts of climate change, and should thus be designed for sustainability and conservation of valuable water resources. For instance, schemes which capture and store rainwater are a useful complement to schemes reliant on accessing groundwater. Schemes should be adaptable in implementation timing and scaled up (or down) in light of the changing nature of water demand.

In a world where water scarcity is very real, demand management is a necessary component of resilient service provision on both the supply and user sides of the process. Efficiencies need to be built across the system, including the need for the water supply industry to reduce leakages. Improving the efficiency of water use at a household level and water conservation at household, business and industrial levels are important factors in reducing water demanded, contributing to a more resilient system.

System strengthening activities should take place, particularly in settings where water supply needs to be scaled up to meet demand, within the overall aim of providing sustainable access for all in the context of international and national targets to achieve the SDGs, with a strong focus on marginalized groups. This would assist in building adequate institutional capability so that the required service level can be provided in the most cost-effective manner, and so that institutions are able to focus on routine asset management, improvement plans and supporting service providers to be financially viable. This also includes installing a policy and legal frameworks aimed at addressing present and future risks and meeting the needs of the public, prioritizing the most vulnerable. The financial viability of services must also be a central consideration, taking into account the ability of consumers to pay for the service and the cost level. Implicit in the concept of system strengthening is the process of building accountability, with all stakeholders (governments, service providers and regulators) discharging their responsibilities in an effective and transparent manner and service users consuming water efficiently. Without accountability throughout the water supply and consumption process, there will be less likelihood of the necessary continuous and systematic improvements.

Finally, a focus on monitoring, learning and capacity building needs to become ingrained throughout the water supply system. This includes enabling effective monitoring of resilience throughout the water supply process, as well as data collection to allow monitoring of implemented changes and to measure their effectiveness and the need for adjustment to maximize impact. Data collection and continuous learning need to take place so that routine corrective action can be taken based on evidence. This process should take place with an understanding of the particular risk and threats that would impact on the water supply system. At a minimum, data should include: hydrometric performance, service levels, customer satisfaction and life-cycle costs. Knowledge should be jointly developed and shared openly so that the best options are implemented, and all stakeholders can buy-in at all stages of the improvement process through building a common understanding, lessons learned from any previous service disruption and improvements required for processes moving forward. Without this data collection, monitoring and feedback, service levels are very likely to gradually decline. When crises occur, then wider support systems also need to be mobilized. National and local institutions may have to take on joint management responsibility. In the absence of these foundations for water supply system resilience, the prospect of *building back better* will be severely limited.

Many of these concepts, practices and approaches are described in practice in the chapters of this book, as described below.

1.8 STRUCTURE OF THIS BOOK

In addition to this introductory chapter, this book consists of eight chapters contributed by different authors working in a range of locations and contexts. The contributors have described how their actions are making water institutions apply resilience in practice, addressing the multiple challenges they face. The result is a collection of standalone case studies, but with cross-cutting themes. Some examples highlight the huge challenges service providers face when working in some of the world's most austere environments. Others focus on very specific challenges and the change people are trying to make to deliver resilient water supplies in industrialized locations.

Each chapter describes the context and the specific resilience challenge that frontline practitioners are facing. They provide examples of the corresponding actions and measures being taken to keep services working efficiently. Although institutions may not actively refer to resilience, they may be nevertheless required to adapt, innovate and learn on an ongoing basis. Each chapter aims to provide frontline experiences, as well as highly practical lessons learnt.

Experiences of how institutions in low- and middle-income countries apply resilience in practice are rarely captured, though they generate some important real-world learning to complement the theory and principles relating to water resilience. Experiences from industrialized countries also offer key insights into how resilience has been built into water supply, incorporating a longer-term vision for sustainable water management. The water sector can be categorized in a range of ways (urban/rural, water/wastewater, developing/industrialized contexts), providing the opportunity to draw out cross-cutting issues, learning from what has worked and avoiding what hasn't. This way of looking at frontline experience also provides the opportunity to understand key principles in building resilient water supplies and which aspects of these systems are critical for the success of all resilient systems.

Chapter 2 is written by Dr. Gisela Kaiser and describes the response to the 'Day Zero' crisis in Cape Town from a public, urban, water supplier management perspective. This chapter details the lessons gained from the period in which the city's water supplies were running drastically low, leading to the implementation of intensive demand management and a long-term plan to diversify water supply.

In Chapter 3, Dr. St John Day, Nitin Jain Tom Menjor and Maada K Penge describe ongoing efforts by Guma Valley Water Company to improve the performance of water supply services in Freetown, Sierra Leone. The difficulties of infrastructure and institutional reform are set out following a decade-long civil war and years of underinvestment. They argue that opportunities to invest in critical infrastructure may have been missed. They also describe the key institutional reforms the water company is pursuing to improve service levels and build resilience in the face of growing environmental, social and economic pressures.

Chapter 4 illustrates a widespread domestic demand management campaign undertaken by a water utility in a rural, peri-urban area of Eastern/Central UK. This chapter is written from the perspective of Essex and Suffolk Water (the water utility company), by Dr. Fatima Aja, Tim Wagstaff and Dr. Liz Sharp.

Chapter 5 again looks at the Eastern Region of the UK, where an innovative and sustainable multi-stakeholder platform known as Water Resources East (WRE), consisting of 130 members representing all major water uses in the region, enables collaborative efforts towards large-scale scenario planning and resilient approaches to water management. The chapter is written by members of the WRE team, Nancy Smith, Dr. Robin Price and Dr. Steve Moncaster.

In Chapter 6, Dr. St John Day and Khaled Mokhtar describe efforts over a five-year period to introduce integrated water resource management approaches at catchment level in eastern Sudan. Khaled highlights the importance of building resilient water supply infrastructure in an arid and water-stressed environment, particularly when water provision institutions have suffered decades of under-financing and neglect. The chapter goes on to describe areas of support required from government, donors and service providers in order to build resilience into future water provision systems.

In Chapter 7, Dr. Ryan Schweitzer, Dr. St John Day, David Githiri Njoroge and Tim Forster ask can and should refugees expect water supply services in protracted emergencies to achieve higher performance levels and resilience. The chapter describes the possibilities for local institutions to be able to manage, modify and finance water services in refugee camps, and explores the efforts needed to raise performance levels in situations of long-term displacement.

Chapter 8, written by Dr. Mohammed Al-Khateeb and Dr. Ali Al-Khateeb, explores the Northern Governates (districts) of Iraq, a fragile and conflict-affected rural area in which the impacts of climate change, governance problems and conflict have led to a programme of augmenting water supply through renewable energy.

In Chapter 9, Orkhan Aliyev and Tim Forster describe how economic approaches to resilience through market-based water provision and systems strengthening have improved service delivery in Tajikistan. This chapter also highlights the remaining gaps as the multi-year programme implemented by Oxfam and its partners draws to a close.

REFERENCES

Abrams L., Palmer I. and Hart T. (2000). Sustainability Management Guidelines. Prepared for Department of Water Affairs and Forestry, South Africa.

Bevan J. (2019). Transcript of a speech, UK Environment Agency. Available at https://www.gov.uk/government/speeches/escaping-the-jaws-of-death-ensuring-enough-water-in-2050 (accessed 30 September 2020)

(UK) Department for Food and Rural Affairs (DEFRA), Environment Agency, DWI, Ofwat (2020). Green Economic Recovery – The Water Industry's Role in Building a Resilient Future. Published letter: Available at https://assets.publishing.service.gov.uk/government/uploads/system/uploads/attachment_data/file/902487/green-recovery-letter-to-water-companies-200720.pdf?utm_source=miragenews&utm_medium=mirage news&utm_campaign=news (accessed 15 April 2021)

GIZ (Deutsche Gesellschaft für Internationale Zusammenarbeit (GIZ) GmbH) (2020). Stop Floating, Start Swimming, GIZ GmbH, Adelphi, Postdam Institute for Climate Impact Research. GIZ, Bonn, Germany.

Hofste R., Reig P. and Schleifer L. (2019). 17 Countries, Home to One-Quarter of the World's Population, Face Extremely High Water Stress. World Resources Institute, Washington, DC, USA.

Howard G., Calow R., Macdonald A. and Bartram J. (2016). Climate change and water and sanitation: likely impacts and emerging trends for action. *Annual Review of Environment and Resources*, **41**, 253–276.

Institution of Civil Engineers (ICE) (2018). Institution of Civil Engineers Professional Skills Review 2018. ICE, London, UK.

Krueger E. H., Borchardt D., Jawitz J. W., Klammler S., Yang S., Zischg J. and Rao P.S.C. (2019). Resilience dynamics of urban water supply security and potential of tipping points. *Earth's Future*, **7**(10), 1167–1191.

The City of Cape Town (2018). Water Outlook 2018 Report: Revision 25. City of Cape Town, South Africa.

The World Bank (2018). Resilient Water Supply and Sanitation Services: The Case of Japan. The World Bank: Washington, USA.

UNICEF (2021). Programmatic Approaches to Water Scarcity, UNICEF Guidance Note, UNICEF, New York, USA.

United Nations Economic and Social Council (2018). Sustainable cities, human mobility and international migration: Report of the Secretary-General, Commission on Population and Development, 51st session, E/CN.9/2018.2. United Nations Economic and Social Council, New York, USA. Available at https://documents-dds-ny.un.org/doc/UNDOC/GEN/N18/024/09/PDF/N1802409.pdf?OpenElement (accessed 30 September 2020)

WaterAid (2011). WaterAid Sustainability Framework. WaterAid UK, London, UK.

Chapter 2

Building water resilience into strategy: The Cape Town drought

Gisela Kaiser

Water Globe Consultants, Cape Town, South Africa

ABSTRACT

The Cape Town drought captured the world's attention at the beginning of 2018 with the announcement of Day Zero: the day that Cape Town's taps would run dry. In the eye of the storm a host of factors contributed to the panic, and rapidly falling dam levels. Politics was exceptionally conflictual, interaction between spheres of government responsible for various aspects of water supply far from perfect, with public perception and media frenzy driving a focus on matters which played a very small part in the effort not to run out of water. During this time, Cape Town was building a water strategy, specifically aimed at making Cape Town more resilient against future droughts by addressing all possible factors contributing to the drought crisis. With dams close to overflowing in 2020, the next challenge is to ensure that the strategy is implemented according to plan.

Keywords: Water scarcity, water shortages, urban water supply, drought strategy.

2.1 INTRODUCTION

Regions reliant on water supply from rainfed dams have always been vulncrable to the impact of drought. This is exacerbated by the uncertainty of future rainfall which is never guaranteed, and reliance is placed on modelling using historic data. While weather has always been variable, climate has been generally reliable. With

doi: 10.2166/9781789061628_0019

anthropogenic activity causing changes in climate, the validity of modelling based on history is currently not fully trusted. Unless the storage capacity is sufficient to carry through numerous seasons of poor rainfall, even with water restrictions to match demand and supply in times of depleted rainfall, the risk of reservoirs running dry remains a threat.

Cape Town experienced its worst drought on record, with climate statisticians estimating a return period of one in 590 years for the low rainfall between 2015 and 2017. The regional water supply system was designed to operate at 98% reliability, resulting in urban users theoretically accepting the risk of severe water restrictions once every 50 years. Agricultural allocations from the regional water supply system are at a lower level of assurance, reflected in lower pricing and higher restrictions being implemented. Despite the levels of assurance having been deemed an acceptable risk within hydrological bulk water planning, how the drought played out proved the system to be less than sufficiently resilient given the uncertainty of the impact climate change.

Through the regional supply system, Cape Town had been served almost exclusively by surface water, which is reliant on runoff from rainfall. Part of the plan to achieve greater resilience thus had to include diversification of water resources away from surface water:

- Groundwater supply is more resilient to drought, although many aquifers need to be recharged to avoid wells running dry and thus even if lagging surface water, groundwater is not fully resilient to drought.
- Water re-use results in better valuing of water by maximising retention of the volume of treated water in the demand cycle. In water scarce regions, flushing potable water down the toilet seems unconscionable, and re-using wastewater, either as treated effluent or directly potable requires consideration. To be able to re-use water, there first has to be sufficient water in the system, thus re-use is also not fully drought-proof.
- Desalinated water is virtually limitless in volume and independent of rainfall. Approximately 97% of the earth's water is saltwater found in oceans and seas. Whilst more expensive, desalination provides supply assurance irrespective of rainfall and drought.

Water supply augmentation is necessary to bring into balance demand and supply, but a number of additional focus areas are critical to render a system resilient. Water scarcity evokes strong emotion, and management of water resources is prone to conflict and politicisation. The protection of rights to water of the most vulnerable is crucial and has historically often been overlooked. Even where relationships are formalised through legislation and regulation, in times of crisis extraordinary collaboration is generally required, spanning all sectors of society. Furthermore, to fully internalise the possible impact of climate change, a long-term view must be taken to having a new relationship with water supply, which inevitably will result in the increased cost of water.

2.2 CONTEXT

South Africa falls into the World Bank categorisation of upper middle-income, based on a per capita gross national income (GNI) range of USD 3996−12,375. The per capita GNI for South Africa in 2019 was calculated at USD 6040, 96th in the world (World Bank, 2020). While inequality has improved slightly since peaking in 2005, South Africa retains the unfortunate top position in the world, with a Gini coefficient (the Gini coefficient is a statistical measure of the degree of variation of income inequality where 0 signifies perfect parity and 1 expresses maximum inequality) of 0.63.

By population, Cape Town is the second largest metropolitan municipality in South Africa, after Johannesburg and covers an area of nearly 2500 km^2. Cape Town is also second to Johannesburg in high net wealth held by private persons. Perhaps counterintuitively, Cape Town's Gini coefficient is the lowest of the metros in South Africa, and lower than the national average at 0.61. In terms of the UNs human development index, Cape Town scores higher than other metros in South Africa, at 0.73 (United Nations Development Programme, 2020).

The City of Cape Town is funded mainly from own revenue from rates (taxes) and utility tariffs, as well as grant funding from the National Government. Rates and tariffs are structured so that higher income households subsidise lower income households that cannot afford to pay. Property values in Cape Town are high compared to that of the rest of the country, and the basic property value used as cut-off for blanket indigency is revisited frequently. In 2020 that value was set at R 300,000 while the monthly household income to qualify as indigent was R 7000. Approximately 40% of households in Cape Town are indigent and thus subsidised by revenue from the balance of households, commerce and industry. These indigent households do not pay for an allocation of 10.5 kl/month or the related sanitation costs. For a household of four, this translates to 87 litres per capita per day (lcd) delivered free of charge. Water tariffs are ring-fenced and cost reflective, and annual increases have historically been kept artificially low, leading to erosion of service standards. In recent years increases have been higher and services are provided at closer to actual cost.

Cape Town has one of the highest levels of biodiversity worldwide. Urban sprawl and high population growth has resulted in encroachment of many ecosystems resulting in threatening many plant species. In response, the City of Cape Town has been at the leading edge of environmental protection for many years, often creating conflict between environmental protection and human habitation. Land invasions typically result in sensitive ecosystems such as wetlands being occupied by informal structures. Once communities are settled in place, eviction processes are lengthy and often unsuccessful, resulting in permanent biodiversity loss.

With a myriad of socio-economic problems in South African cities, there is always a tension in competing for scarce resources. Trade-offs therefore need to be made at every level; government does not have the financial resources to fund all the infrastructure required countrywide. In 2020 in South Africa, it is common cause that all utilities and shared infrastructure in most regions are in need of investment not only to provide new infrastructure but also to maintain that which exists. In most metropolitan municipalities road infrastructure is crumbling, and rates income cannot afford replacing extensive road networks. The national electricity grid is under supplied due to decades of poor maintenance or renewal, and lengthy delays in commissioning of new power plants further hampered by momentous corruption. As a result, the country is subjected to frequent periods of load-shedding where load is reduced through sequentially switching off areas for a couple of hours at a time. Solid waste services are woefully under-resourced, with many rural municipalities having virtually no capacity to manage and dispose of waste. As a result, illegal dumping abounds with detrimental consequences on the environment. Similarly, the condition of both bulk and reticulation water and sanitation infrastructure has been eroded over decades, while new infrastructure has been targeted at providing access to those most in need. With an estimated one in eight residents of Cape Town living in informal settlements, in need of affordable, well-located housing opportunities, the total lack of resilience in the infrastructure space is unmistakeable.

The water supply system serving Cape Town has had a 98% level of assurance, and compared to other services, was generally deemed to be sufficiently resilient prior to 2017. Water supply was reliable, at good quality, and droughts were managed by imposing restrictions while local government worked with the National Department of Water and Sanitation (NDWS) on reconciliation of the system. By the middle of 2017, the City's political leadership assessed the risk of climate change accelerating faster than compensating infrastructure could be built to be unacceptably high. Water, being such a core component of *all* life, demands special priority in building resilience, and with full focus on the drought, water held centre stage in the Western Cape for many months.

Worldwide, infrastructure development programmes have for decades been seen as the path to economic growth and social upliftment, and since 1994, the South African government has embarked on numerous programmes to do exactly that. From the Reconstruction and Development Programme (RDP), the Growth, Employment and Redistribution strategy (GEAR) through the Accelerated and Shared Growth Initiative for South Africa (ASGISA), New Growth Path (NGP), National Development Plan (NDP), and in 2020, South Africa's Economic Reconstruction and Recovery plan developed to counter the disastrous economic impact of the extended coronavirus disease 2019 lockdown.

Despite the series of laudable customised programmes, poverty, high unemployment, low economic growth and austere inequality prevails in South Africa. An unfortunate culture of corruption resulted in severe erosion of the country's development, and only in 2020 are perpetrators finally being brought to book. The damage done is incalculable, but the past cannot be changed and the country's best chance at moving forward is to transform the culture to one where rules are respected, and to do so without delay. The new economic recovery plan not only aims to invest in infrastructure thereby stimulating the economy and creating jobs, but also to fight corruption and build a capable state. Furthermore, the intent is clear that obstacles to private sector-led employment and economic growth need to be removed with a focus on local production and consumption to stimulate industry. Planning and budgeting for new bulk water infrastructure is key to the country's water security, and being recognised at the highest level, there is hope that implementation will follow and value for money achieved through the elimination of corruption.

Over and above the risk to resilience posed by degraded infrastructure in the country, relying solely on rainfed dams adds significantly to the risk of inadequate water supply. With rainfall becoming less reliable over time, consideration must be given to diversifying water supply sources.

2.3 INSTITUTIONS

Democratic South Africa is a parliamentary republic and has three spheres of government – national, provincial and local. The South African Constitution (Act 108 of 1996), prescribes that everyone has a right to have access to sufficient water, amongst other basic needs, and that the state has responsibility to progressively realise these rights (Republic of South Africa, 1996). NDWS is responsible for formulation and implementation of policy governing water resources management, which covers protecting, managing, using, developing, conserving and controlling the country's water resources. In terms of the National Water Act, Act 36 of 1998, National Government is responsible for and has authority over water resources, including allocation, protection and use of water (Republic of South Africa, 1998).

Provincial government has no direct responsibility in provision of water, but as a minimum practice oversight through a number of departments. These include:

- The Department of Local Government aims to support and strengthen the capacity of municipalities to manage their own affairs. This department is also responsible for disaster management, and as such participated actively subsequent to the local and provincial disaster declarations.
- The Department of Agriculture stimulates regional economic growth through research and support to the agricultural sector, a major water user in the province. With extreme restrictions on water during the drought, support was required in managing job losses and improved efficiency in agricultural practices.
- The Department of Economic Development and Tourism visions a vibrant economy in the Province and supports business and industry to grow the economy and employment. The threat of literally running out of water

alarmed many businesses to halt expansion plans and invest in alternative water supplies and efficiency measures.
- The Department of Environmental Affairs and Development Planning safeguards the natural environment as custodian of various approval processes while supporting sustainable development through regulating development. Their role in the drought was amplified through approvals related to new augmentation schemes in terms of emergency drought conditions.

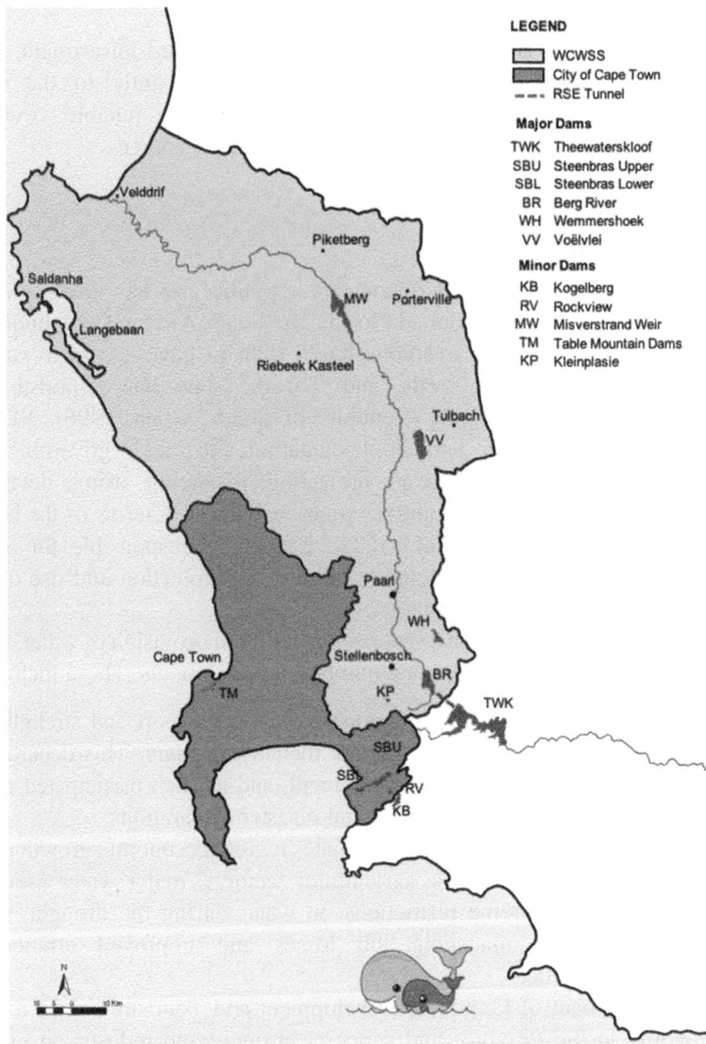

Figure 2.1 The WCWSS and dams.

In terms of the Water Services Act, Act 108 of 1997 municipalities are defined as Water Services Authorities, responsible for ensuring access to water services (Republic of South Africa, 1997). Part B of Schedule 4 to the Constitution lists water and sanitation services limited to potable water supply systems and domestic wastewater and sewage disposal systems as a local government matter. The City of Cape Town's Water By-law defines the terms of provision of water within the municipal area (City of Cape Town, 2010).

Whereas most municipalities in South Africa purchase bulk water from water boards, the City of Cape Town is also responsible for provision and treatment of bulk water, and owns three of the six dams which comprise the Western Cape Water Supply System (WCWSS). The WCWSS supplies the urban demand of Cape Town as well as several local and district municipalities, agricultural users and irrigation boards. Cape Town annually utilises around 64% of water from the system, with agriculture using 29% and smaller municipalities making up the balance.

The extent of the WCWSS is shown in Figure 2.1. Most of Cape Town's water comes from the Riviersonderend-Berg River Water Scheme, which makes up the biggest part of WCWSS. This scheme captures the flow of three main rivers: the Sonderend River, Berg River and Eerste River, which feed into six major dams. The long-term sustainability of water resources in the WCWSS is achieved through the provision of flexibility of City infrastructure to shift the demand among various water sources within the system, to the benefit of all users.

Over and above legislative prescripts amongst different spheres of government, the relationship between the City and the NDWS in managing the system collaboratively is defined in the Raw Water Supply Agreement. This agreement was finalised to facilitate funding of the last large dam to be constructed in the system, that being the Berg River dam. The system is premised on operating rules, which when adhered to results in the system being in balance, with assurance of supply reconciled with growth in demand. The operating rules are

2.4 MINIMISING SPILLAGE

As Cape Town is the largest water user consistently throughout the year, the City's demand is shifted to those dams more likely to spill during winter, while other dams still have storage capacity available. The City provided additional treatment capacity at its water treatment plants, and spare capacity in conveyance infrastructure to provide this flexibility. While it is beneficial for the City to utilise water from its own dams thereby avoiding pumping costs for water transfer, by shifting the demand to any of the dams in the system, yield can be optimised.

2.5 MINIMISING WASTAGE

Through the City's Water Conservation and Water Demand Management (WCWDM) programme, the City monitors monthly urban demand and implements strategies to continuously reduce water wastage. Similarly, the agricultural sector has management structures in place to manage water use, and the responsibility for managing and enforcing responsible use falls on the NDWS.

2.6 RESTRICTING DEMAND

At the end of the hydrological year on 31 October every year, the system is evaluated to determine whether restrictions need to be implemented for the coming year. The evaluation model considers available storage, demand and forecasts different inflow scenarios to determine the appropriate level of restriction, if any, to be imposed.

The WCWSS is represented by urban and agricultural water users as well as the NDWS in overseeing the supply and demand reconciliation strategy. The strategy was first developed in 2007, and thereafter updated regularly with actual demand and supply statistics.

Located in a winter rainfall region, around 90% of runoff occurs between May and October. During the balance of the year, November to April, around 70% of water is used: urban demand varies only marginally over the year while agricultural use occurs mainly over the summer months. The system was designed to rely on reasonable rainfall every winter, with average runoff sufficient to fill all but the largest dam every year. The storage capacity of the largest dam, Theewaterskloof, is such that approximately 2 years' of average rainfall is required to fill it from scratch, thus 50% of storage capacity provides carry-over in times of drought.

On the supply side, the impact of climate change, including reduced rainfall and increased evaporation was considered as one of the scenarios, together with, for example, the reduction in yield due to spread invasive alien vegetation. The eradication of these species has been problematic as much of the catchment areas are difficult to access, and land ownership both private and in the hands of various different state departments in the different spheres of the government. Even with programmes such as the much lauded Working for Water under the Department of Environmental Affairs which started in 1995, where job creation went hand-in-hand with catchment management, progress has deteriorated and the calculated impact of the reduction in yield during the drought had a significantly negative impact on dam levels.

Legislation in South Africa is progressive and comprehensive, and governance is highly regulated. Despite the complicated responsibilities assigned to the three spheres of government, there does not seem to be a policy deficit in the ambit of bulk water management which aggravated the impact of the drought. Rather, common to South Africa is the ability to remain on-track with strategy implementation. This is compounded by politics negatively impacting with frequent changes in leadership at ministerial level, which inevitably results in new strategies and implementation plans being developed, budget re-prioritisation and changes in professional staff especially in leadership positions. Combined, these not only delay implementation but also reduce value achieved in the continuous re-drafting of plans and strategies.

2.7 TIMELINE

The capacity of the major dams in the WCWSS is close to 900 million cubic meters (MCM). The unrestricted urban allocation in 2016 was 415 MCM, and agricultural allocation nearly 200 MCM, comprising 68% of stored capacity. With unrestricted demand, capacity can provide sufficient water for only one and a half years. Restricted by 25%, the storage volume can supply demand for nearly 2 years.

From the supply side, with average runoff from rainfall of 711 MCM, dams would spill at a starting level after summer of 20%. Average rainfall is thus more than sufficient to provide secure supply – between 2000 and 2015, dams spilt eight times. Slightly higher than average rainfall in the three years preceding the drought had dams spilling in the winters of 2012, 2013 and 2014.

The 2015 winter rainfall season heralded the first year of the drought (see Figure 2.2). With the highest urban demand in 15 years, dam levels fell from full to below 50%. The low rainfall in 2015 resulted in dams recovering to only 74% at the end of the rainy season. With Cape Town restricting demand by 10%, reduction in dam levels was ~7% less than the previous year, and reached a minimum average of 29% before winter 2016. Rainfall and thus runoff in 2016 was slightly better, resulting in dam levels rising by nearly 33%. With harsher

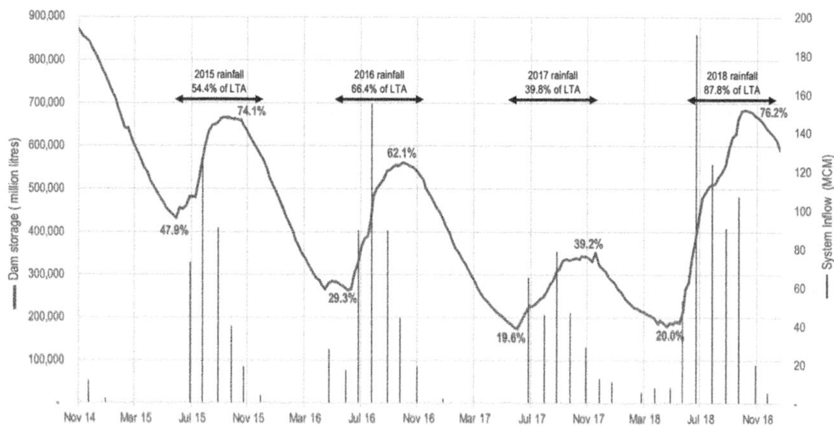

Figure 2.2 WCWSS dam levels and seasonal inflow 2015–2018.

restrictions in place (but not imposed) in the summer of 2016/2017, dam levels reduced by more than 40%, ending just below 20% full. Lowest rainfall on record in 2017 resulted in dam levels peaking just below 40% by the end of the rainy season. The impact of the disaster declaration and various restriction measures is evident in a reduction of less than 20% in the summer of 2017/2018. Due to the dramatically reduced demand, lower-than-average rainfall in 2018 had dam levels rise to 76%, an increase of 56%.

Much happened in the City administration during the period depicted in Figure 2.2, creating significant political instability, whilst the impact of the drought increased to a peak late in 2017 and only started to calm down towards the end of 2018. Local government elections were held in August 2016 and the new city leadership, with a restructured administration was appointed from January 2017. In January 2018, oversight of the drought was moved from the Mayor to the Deputy Mayor. A number of votes of no-confidence failed early in 2018 before the Mayor eventually reached an agreement to step down and leave the ruling political party.

2.8 DESCRIPTION OF ACTIVITIES

The strategy to practically navigate the drought had three main components: managing dam storage in reducing drawdown as far as possible, managing Cape Town's demand, and accelerating sources of additional water.

2.8.1 Managing dam storage

The first element was to retain the remaining water in the dams for as long as possible. Cape Town gets its water from the WCWSS, which is controlled by the NDWS. During the previous summer months of 2016/2017, both urban and agricultural users were restricted by 20%. Urban users made a saving of 17% while agriculture used 3% *more* than their unrestricted allocation. Figure 2.3 shows the planned demand for both Cape Town and agriculture for the 2017 hydrological year, with a 45% urban and 55% agricultural restriction.

The graphs in Figure 2.3 were made public by the City to ensure broad awareness of why dam levels were dropping so rapidly in summer 2017/2018. Urban demand is shown on the left. The usage target was established by reducing the City's allocation by 45%, to just below 175 MCM. By the end of January 2018, the City was tracking marginally above planned demand. Releases to agriculture started earlier than they should have at the end of 2017, and continued at the rate that the 55% restricted target of 58.2 MCM would have been exceeded by February 2018 whereas usually allocations were made through April. As Cape Town had no control on physically stopping releases to agriculture, the City repeatedly alerted NDWS to the need to stay within the restrictions required by the operating rules.

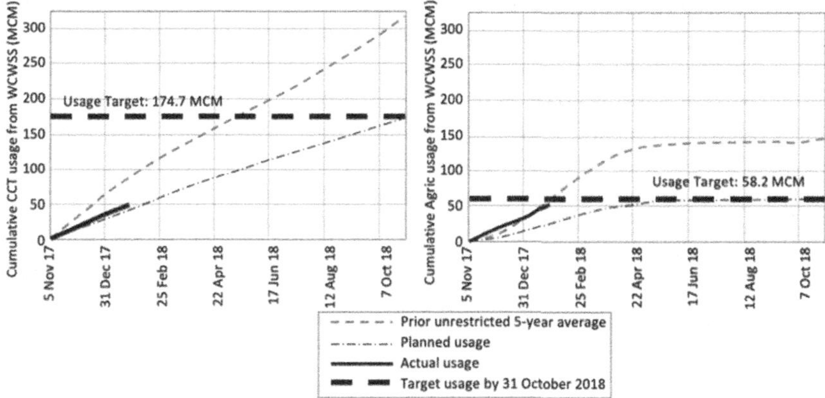

Figure 2.3 Demand tracking to January 2018.

2.8.2 Managing demand

The second component of navigating the drought managed by the City of Cape Town was to ensure that our urban demand was reduced to an absolute minimum to use less water from the system. Demand had already reduced dramatically by the end of 2017, and with an increase in tariff and threat of Day Zero in January 2018, the daily demand finally broke the demand threshold of less than 500 million litres per day (MLD). This was a significant achievement as peak summer demand was in the region of 1200 MLD. Cape Town was widely recognised for the significant achievement, such as a certificate of excellence received from the International Water Association for reducing demand by 55% without resorting to intermittent supply.

As a standard response to drought, restrictions had been imposed in the region since water records were kept for Cape Town since around 1834. Reasons for restrictions historically included inadequate water resources to meet demand, droughts and insufficient infrastructure to treat or convey water. Often, restrictions had to be imposed in the periods immediately prior to completion of major infrastructure such as Wemmershoek Dam, Voelvlei Dam and Faure water treatment plant. Restrictions due to drought were most recently imposed in 2000 and 2004. Between 2004 and 2014, dam levels always exceeded 85% by the end of winter, and exceeded 100% in 6 of the 10 years.

Reducing demand so dramatically in the years of 2016/2017/2018 relied on a multi-pronged strategy:

(1) **Communications:** To engage more than 4 million people in saving water required simplifying complex messages to have broad appeal. All media channels were employed, from print, to radio interviews through to personal interactions to targeting information at the right level of detail in an attempt to soothe the tumultuous political environment.

(2) **Pressure management:** The existing pressure management project was accelerated during the drought. Division of the city into hundreds of discrete pressure zones was finalised, to allow for management of pressure within each zone to be automatically regulated and remotely controlled.

(3) **Restrictions and punitive tariffs:** Water restrictions were increased progressively as the drought unfolded. At the start, three restriction levels were defined, each with behavioural restrictions and tariffs to ensure revenue balancing on reduced demand. This was increased to seven restriction levels by the end of the drought, with each restriction level defining six steps of volumetric use. Only at level 6 were tariffs sufficiently punitive above 6 kilolitres per household per month to result in decreased usage.

(4) **Flow limiters:** Through its indigent support programme, which was first implemented in 2007, the City repaired household leaks and installed a water management device at households which could not afford to pay for water. The device provides 10.5 kilolitres per household per month, as the meter physically restricts water supply to 350 litres per day. During the drought, such meters were installed to limit flow at delinquent households with excessive consumption. These households were charged for the cost of the meter as well as installation. Over 50,000 flow restrictors were installed in the summer of 2017/2018.

(5) **Adaptation measures:** Demand was further reduced by a number of citywide and household adaptation measures, including increased use of treated effluent, retrofitting public buildings with water-efficient fittings, encouraging rainwater harvesting and greywater usage at private households, water-sensitive urban design and so on.

2.8.3 Accelerating augmentation

The final component of navigating the drought was to accelerate additional water from diverse augmentation schemes into the system. It was recognised that these would not be sufficient to significantly increase supplies in the short-term, but the urgency with which projects were undertaken was to set the tone for the augmentation schemes which were later approved in the water strategy. The city had been working on groundwater, potable re-use and desalination projects for decades, which provided a sound base for the accelerated work.

It is accepted in engineering practice that it is seldom feasible to build one's way out of a drought. Water augmentation projects at scale are well regulated and complex, and regionally in the Western Cape, virtually all development and management experience was from surface water. The scale of the volume of water to be produced also means that small inefficiencies can have a significant impact on cost. Therefore, procuring to obtain best value for money while meeting specifications is critically important.

Groundwater projects were expected to be quicker to implement to first water than either desalination or direct re-use. Infrastructure for groundwater projects are naturally more geographically spread out than dams for example, and easier to accommodate in existing reticulation schemes. The Cape Flats aquifer is a shallow sandy aquifer, located within the urban area and was prioritised to be the quickest project for additional water. A variety of delays meant that first water into the system was only available after the worst of the drought. Delays were due to, inter alia, lower than expected borehole flow-rates, poorer than expected water quality, limited accessibility to suitable sites, vandalism of construction site establishment and environmental objections.

Exploratory and pilot drilling work on the Table Mountain Group aquifer project had been underway since 2003. The best areas of accessibility and anticipated yield is located in environmentally protected areas, many of which are close to existing dams. Although the drought disaster declaration excluded these projects from requiring full environmental authorisation, the City followed a conservative approach to ensure long-term sustainability and optimisation of the volumes of water to be finally extracted from this aquifer. The project provided the first *new* water into the WCWSS from production boreholes drilled at Steenbras Dam. Water from the aquifer is of good quality requiring treatment only to remove iron and manganese.

Desalination contributed half of the target of 500 MLD of the initial drought response, in three projects which were fortunately never implemented: 50 MLD from a fast-tracked land-based plant and 200 MLD from ships and barges. A host of small-scale temporary containerised desalination and wastewater re-use projects were initiated in mid-2017 with a total yield just over 100 MLD. Of these, three desalination plants with a combined yield of 16 MLD and one re-use plant with 10 MLD capacity were implemented. The desalination projects provided water in May 2018, as winter rain started falling. The re-use plant suffered some delays and provided water only a year later, at which point the social acceptability concern resulted in water not being injected in the reticulation system. Instead, water was used for utility purposes on site, and as collectable water for purposes such as street cleaning. The cost of water from the temporary plants varied between R 30 and R 35/m^3 compared to that from surface water of R 5.20/m^3 at the time.

As work progressed on the three-pronged approach to navigate the drought, a strategy was taking shape towards building resilience. This evolved over time, but as a starting point, the need for a comprehensive water strategy was built upon a combination of the following objectives:

- Don't lose the lessons learnt during the drought! Value water. Build better social cohesion and equity in access.
- Consider, plan and mitigate shocks (drought, tariff increase, localised flooding, storm surge, protest action) and stresses (inward migration,

informal settlements, poor hygiene and sanitation, sub-optimal institutions, ageing infrastructure).

- Diversify supply by reducing reliance on rainfall; introduce redundancy.
- Maintain water conservation and demand management, support household resilience.
- Resolve better management of catchments requiring cooperation between all spheres of government.
- Price water appropriately.
- Entrench water-sensitive design by managing the entire urban water cycle.

2.9 ANALYSIS OF IMPACT

While the drought was underway, it was recognised that a more formalised strategy was required to ensure long-term water resilience. At the time, Cape Town was also developing a broad municipal resilience strategy. This strategy defines resilience as 'the capacity of individuals, communities, institutions, businesses and systems in a city to survive, adapt and thrive no matter what kind of chronic stresses and acute shocks they experience' (City of Cape Town, 2019a). Chronic stresses include those occurring perpetually such as food insecurity, high unemployment and substance abuse, which weaken society continuously. Droughts are included under acute shocks, with sudden events such as fires, floods and pandemics.

Through the water strategy, Cape Town has a vision of becoming a water-sensitive city by 2040, where management of water resources are optimised and integrated to improve resilience to the benefit of all people in the city (International Water Association, 2019).

Cape Town's Water Strategy is built on five commitments, which if implemented, will result in Cape Town being more resilient to future droughts which will inevitably occur (City of Cape Town, 2019b). As a starting point the City wanted to prioritise rebuilding of trust as this was very much eroded in misinformation and political arguments in the media during the drought. For this reason, the word 'commitment' was chosen to underscore the administration's willingness to give time and energy to the promise of building a water-resilient city.

Ten principles guided the content of the strategy, emphasising that a new relationship with water is required. Water has historically been under-priced, and in times of abundance, not valued. With a focus on people, the strategy recognises that people have different requirements and cultural needs in their relationship with water. Inclusivity and trust will be built by transparency in engaging broadly with stakeholders across boundaries and party politics. Utilising water to create natural connections with nature for the enjoyment of citizens will benefit ecosystems, and build on the knowledge base of local culture.

The commitments provide the backbone of the water strategy and were carefully selected to provide a holistic strategy, covering all the major actions required to build resilience in Cape Town's relationship with water.

2.9.1 Safe access to water and sanitation

Safe access requires municipal potable water supply to meet stringent quality standards, while being available to all inhabitants, business and industry. In unequal societies this is particularly important, where much of the population lives in rudimentary accommodation without direct service connections, and share water points and communal toilets. A commitment was made to engage with poorer communities to develop appropriate service standards that could improve the lived experience, even if not offering permanent formal housing with individual service connections as an option. In the drought, much noise was made about water wastage in informal areas, but in reality, 15% of people in the city accounted for less than 5% of water usage. Decreasing social inequality and prioritising the dignity of all people needs to be a major consideration for all cities.

2.9.2 Wise use

With the dramatic reduction in consumption achieved by 2018, it was evident that the average person could use much less water than before 2015. Rather than demand bouncing back to pre-drought levels, demand should remain supressed, recognising that water must be respected and used wisely to build future resilience. To attain wise use, in the first instance water needs be priced correctly. Municipalities do not make a profit on sale of utility services, but need to cover the cost of providing services. This means that the cost of additional infrastructure needs to be covered by an increase in tariff, while still maintaining a progressive and affordable tariff regime. The second aspect within a local authority's control to influence wise water use is to amend planning regulations, bye laws and green building incentives. Thirdly, with customer complaints escalating exponentially in the drought, on top of increased tariffs and delays in answering queries, the relationship with the customer base had to be dramatically improved, by adding capacity to the support and back office staff. The final component of wise use is to ensure that infrastructure is well maintained and pressure appropriate to minimise leaks and reduce water losses below Cape Town's achievement of 15%.

2.9.3 Sufficient, reliable water from diverse sources

Building on the augmentation projects identified in the reconciliation strategy and accelerated in the drought, the timing and additional yield required over time, with due consideration of cost, resulted in proposed additional volumes in the finalised water strategy as shown in Table 2.1. Modelling anticipated water demand together with impacts of climate change, such as lower rainfall and more frequent drought, the most likely requirement for new, additional water supply over a 10-year period was calculated to be ~350 MLD. Together with the existing surface water resources in the WCWSS, this provided a level of assurance of 99.5%, equating to serious restrictions being required only once in

Table 2.1 Water strategy committed 10-year build programme.

Intervention[+]	First Water	Effective Yield		Total Capex	Unit Capex[++]	Operating Cost[+++]
		MLD	*MCM pa*	R million	Rm /MLD	R/klitres
Demand management	2019	70	26	410	6	3
Alien vegetation clearing	2019	55	20			~1–2
Management of WCWSS	N/A	27	10			~0.2–0.5
Cape Flats Aquifer P1	2020	20	7.3	610	31	5
Table Mountain Group P1	2020	15	5.5	375	25	5
Cape Flats Aquifer P2	2021	25	9.1	450	18	5
Atlantis Aquifer	2021	10	4	290	29	8
Table Mountain Group P2	2022	15	5.5	335	23	5
Table Mountain Group P3	2022	20	7.3	326	16	2
Berg River Augmentation	2023	40	15			~3–5
Water Re-Use P1	2024	70	26	1360	20	5
Desalination Phase 1	2026	50	18	1650	33–40	9
Total including WDM		417	154	5806		
Total new supply		347	128	5396		

Notes:

[+]Timing, and the capital and operating costs are best available engineering estimates. All schemes subject to outcomes of ongoing investigations (to determine optimal yield, siting and timing) and relevant approvals.

[++]Rounded to nearest million Rand.

[+++]Rounded to nearest Rand.

200 years, translating to a fourfold increase in reliability. Over and above the new supplies, renewed demand management interventions resulted in an additional availability of 70 MLD in the system. This would result in new, and affordable supply to 2030, providing a springboard for further incremental augmentation thereafter.

The system information would be monitored annually, providing for adaptation of the build programme to retain balance should either demand or supply assumptions change. Given the size of the system, dramatic rapid changes that could not be mitigated by adapting the build programme were not anticipated.

2.9.4 Shared benefits from regional water resources

Responsibility for bulk surface water schemes had been the responsibility of the national DWS since the promulgation of the Water Act in 1956. Although the impact of the new water programme had not been resolved with DWS, Cape Town decided to proceed, to improve resilience to anticipated future droughts. Aspects such as existing allocation, level of assurance and cost of raw water had not been resolved. However, to have paused until legal agreements were concluded would have exposed Cape Town to further risk.

Woking in partnership, and collaborating with stakeholders on water was not generally a priority prior to the drought, but during, many valuable relationships were established. It was recognised that virtually everyone could be significantly impacted should water supply run dry, and collaboration was key to protecting the other commitments in the strategy. The intention thus was to build on this level of transparency and collaborate across business, customers, civil society and spheres of government.

2.9.5 A water-sensitive city

Integrated urban planning and design aimed at achieving a water-sensitive city requires integration of the total urban water cycle, which includes bulk water supply, stormwater, groundwater, reticulation networks and wastewater. Infrastructure systems designed purposefully to facilitate water sensitivity provides for *valuing* every part of the water cycle, for example that stormwater is slowed and stored to recharge aquifers rather than being guided out to sea. Much of Cape Town's urban sprawl has occurred in low-lying areas prone to flooding while summer months are hot and dry. Mechanisms for flood control and utilisation of water for aesthetic and recreational purposes will change the landscape, and the commitment is for all city-led development to follow water-sensitive design principles while encouraging private development to do the same.

2.10 CONCLUSION

It can be argued that the region and Cape Town was technically still resilient to multi-year droughts at the time of the drought between 2015 and 2018, but practically, failures in adhering to the operating rules and lack of cooperation had exacerbated the level of drought disaster significantly. It has been calculated that dams would have had an estimated 18% more water had the operating rules been followed. This would have meant that levels were at 56% rather than the actual 38% at the start of summer 2017/2018. Restrictions would still have been required, but the panic around running out of water would have been averted.

Bulk water planning and the demand and supply reconciliation strategy provided a sound base to deal with a multi-year drought at the level that the world has experienced climate change to date. There is no guarantee that future droughts won't be more severe or that rainfall patterns may change significantly and suddenly, and there for the best way to build resilience in water supply is to diversify and provide some redundancy in the system, even if financial resources are limited.

Having a municipal Council-approved strategy in place provides a solid foundation for building resilience. Commitment is required by the utility department to ensure that sufficient resources are made available and that the necessary support is provided to implement the strategy. Only once all the commitments of the Water Strategy have been met will Cape Town be resilient to future droughts.

REFERENCES

City of Cape Town (2010). Water By-Law. Available at https://resource.capetown. gov.za/documentcentre/Documents/Bylaws%20and%20policies/Water%20Bylaw% 202010.pdf (accessed July 2021).

City of Cape Town (2019a). Cape Town Resilience Strategy. Available at https://resource. capetown.gov.za/documentcentre/Documents/city%20strategies%2C%20plans%20and %20frameworks/Resilience_Strategy.pdf (accessed July 2021).

City of Cape Town (2019b). Our shared water future: Cape Town's Water Strategy. Available at https://resource.capetown.gov.za/documentcentre/Documents/City%20strategies, %20plans%20and%20frameworks/Cape%20Town%20Water%20Strategy.pdf (accessed July 2021).

International Water Association (2019). The Source: Realising the vision of a water sensitive city. Available at https://www.thesourcemagazine.org/realising-the-vision-of-a-water-sensitive-city/ (accessed July 2021).

Republic of South Africa (1996). Constitution of the Republic of South Africa No. 108 of 1996. Government Printer, Pretoria.

Republic of South Africa (1997). Water Services Act, No 108 of 1997. Government Printer, Pretoria.

Republic of South Africa (1998). National Water Act, No 36 of 1998. Government Printer, Pretoria.

United Nations Development Programme (2020). Human Development Reports. Available at https://data.worldbank.org/indicator/NY.GNP.PCAP.CD?locations=ZA-XT (accessed July 2021).

World Bank (2020). GNI per capita – South Africa, Upper middle income. Available at http://www.hdr.undp.org/en/countries/profiles/ZAF (accessed July 2021).

Chapter 3

Transforming a water company to improve service levels and resilience: Lessons from Sierra Leone

St John Day[1], Nitin Jain[2], Tom Menjor[3] and Maada K Penge[4]

[1]*Principal Consultant IOD PARC, Edinburgh, UK*
[2]*Infrastructure Resilience Specialist, Global Center on Adaptation, Netherlands*
[3]*Monitoring, Evaluation and Accountability Manager, Guma Valley Water Company, Freetown, Sierra Leone*
[4]*Managing Director, Guma Valley Water Company, Freetown, Sierra Leone*

ABSTRACT

All water companies need to be able to provide safe, adequate and reliable water supplies to their customers and consumers. Yet some work under very daunting conditions. The civil war in Sierra Leone resulted in the destruction of much water supply infrastructure. It also had a devastating impact on the performance of water companies. Since the war ended in 2002 other changes continue, such as: population growth, unplanned urbanisation, environmental destruction and climate change, plus the Ebola outbreak. These pressures all have a massive impact on the natural environment and on demands for water. It is against this background that Guma Valley Water Company is trying to rebuild water infrastructure and strengthen utility arrangements for providing a reliable and affordable service on which people depend. This article describes ongoing efforts to improve water supply in Freetown. The case study highlights the multi-faceted nature of resilience building and the processes that must be undertaken if water companies are to become resilient. Long-term technical and financial support is required, however, programmes should be realistic in their expectations.

Keywords: low-income countries, water utility reform, water resources, infrastructure.

3.1 INTRODUCTION

Despite recent development progress, Sierra Leone is still classed as a fragile state (Fragile states are classified as those that are unable or unwilling to provide essential services to their people and communities). Typically, this means that periods of relative stability, in which development efforts make some progress, are often punctuated by crises that have long-lasting impacts. The 2014 West African Ebola outbreak and 2017 mudslides in Freetown highlight recent disasters.

Sierra Leoneans are resourceful people and Guma Valley Water Company (hereafter referred to as 'Guma') is a stoic company. However, it is a daunting prospect to achieve resilient services if water infrastructure is crumbling and the company struggles to achieve its desired level of performance. It is against this background that Guma must develop both its infrastructure base and own capability to deliver water to the people of Freetown, in order to fulfil its mandate.

The single biggest challenge for Guma, is that physical water demand massively exceeds available supply. There is simply not enough water entering the water supply network to meet the demand side pressures, which continue to grow. This fundamental point has often been overlooked by successive governments and donors that are responsible for designing technical support programmes. Securing new raw water sources represents the single most important area of improvement for Guma's service levels.

This chapter describes both previous efforts, and future requirements, to strengthen water resources management, water supply and company performance. This is required so Guma can achieve its aspiration to 'provide access to safe, affordable, and sustainable water for all residents of Freetown by 2028'. This case study reflects on the transitions undertaken and concludes by sharing key learning from these experiences.

3.2 OVERVIEW

Sierra Leone is located in Western Africa on the Atlantic Coast. It has a tropical climate which is strongly influenced by the West African Monsoon. Its natural environment is characterised as being well-watered. Average annual rainfall ranges from less than 2000 mm in drier areas of the far north east to about 2500 mm in the south east and more than 4500 mm over the Freetown peninsula

(MWR, 2015). However, a substantial amount of precipitation (around half) is lost to evapotranspiration and, in the absence of major abstractions and surface water storage infrastructure (such as reservoirs), the many rivers discharge surface runoff directly to the Atlantic Ocean. Rainfall across Sierra Leone is also seasonal, typically occurring from May to October followed by a protracted dry spell from November to April. Rainfall peaks in July and August and this has a direct impact on Guma's ability to serve people year-round. Most of Guma's rainfall monitoring network along the peninsula was also destroyed during the civil war and there is currently a dearth of routinely collected and analysed hydrometric data to assess seasonal variations year-on-year.

Continuous problems related to water resources management include amongst others: population growth, unplanned urbanisation, catchment encroachment, land degradation, deforestation and growing demand. These issues, along with growing water demands, vandalism to water supply infrastructure, leakage, high rates of non-revenue water, limited revenue and inadequate financial investment over many years is undermining service performance levels and making water supply an issue of national security. As an indication, the Ministry of Water Resources (MWR) and Guma have been operating a drought contingency plan in Freetown during the dry season months since 2016.

Indeed, the performance of the water sector in Sierra Leone has been mixed in recent years. Sierra Leone's economy and institutions suffered massively as a result of the civil war (1991–2002) and the network of infrastructure, institutions and finance still require significant external support nearly two decades later. The performance of service providers has also struggled to keep pace with the many growing pressures and drivers of change.

3.3 CHALLENGES OF WATER RESOURCES MANAGEMENT, WATER SUPPLY AND UTILITY REFORM

The Guma Valley watershed is located on the Freetown peninsula and is one of the wettest in Africa (Ledger, 1975). It drains into the Guma reservoir and dam. This is the main source (Other smaller sources include: Hastings dam, Charlotte, Regent, Babadori, Bluewater and Whitewater.) of water for Freetown and is owned and operated by Guma. The dam was constructed in 1967 when the city's population was around 450,000 people. The original dam, treatment works and trunk mains were commissioned with a capacity of 27,000 m^3 water supply per day. This was supported by further upgrading work in 1973 and subsequent infrastructure works that increased capacity to 80,000 m^3 water supply daily. To compliment this construction work, further efforts were taken to extend water supply to Wellington Industrial area in 1985.

In 2003, the population in Guma's supply area was estimated to be in excess of 825,000. To compound these demand−supply pressures further the network has suffered from ongoing vandalism (illegal connections), destructive cutting of pipes, limited operation and maintenance throughout the civil war and negligible financial investment. The population in Freetown has continued to grow rapidly since 2003 and Guma's supply area now covers an estimated 1.5 million people (Statistics Sierra Leone 2017). The GVWC Water Act (2017) extended Guma's supply area as the entire peninsular and therefore includes Freetown, Waterloo and the peri-urban and rural areas along the peninsula. This includes more than 70 unplanned or informal settlements that have emerged in the past 20 years.

Figure 3.1 shows how Guma's operational area is divided into three supply areas that broadly consist of: urban communities and households in the west, business and trading areas in the central district, and people who live in densely populated eastern Freetown.

Recent work to measure the flow of water from the Guma dam determined the reliable supply is in the order of 60 MLD. This is enough water to supply only 40 lpcd assuming the water can be distributed equally with no losses to a population of 1.5 million people. In truth, losses are considerable and equity

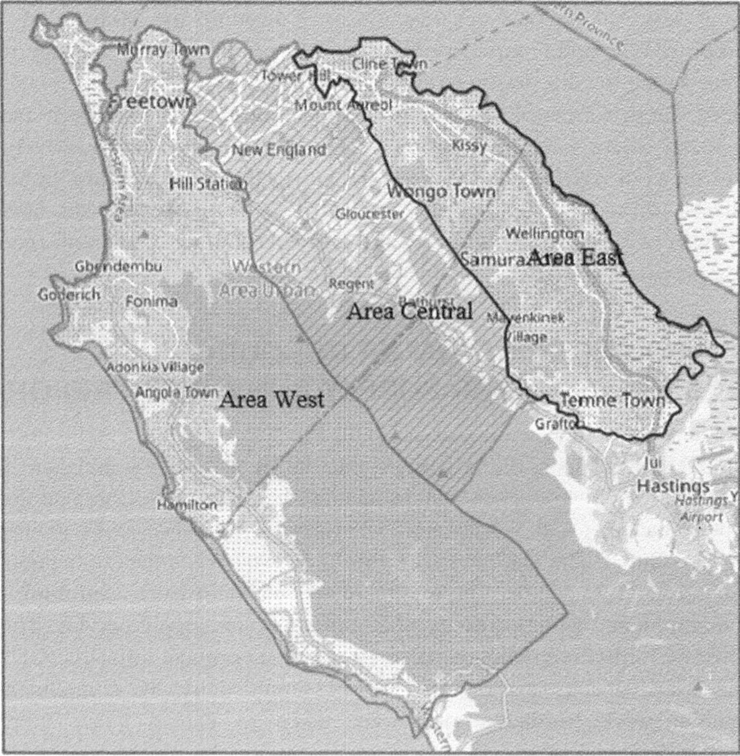

Figure 3.1 Guma valley water company operating area. *Source*: Guma Valley Water Company (2019), Business and Investment Plan; Available from Guma Valley Water Company, Freetown, Sierra Leone.

across the city is a major challenge. New raw water sources have to be protected, harnessed and developed, however, small local water sources are threatened by urbanisation that is damaging all potential catchment areas. Many hydrometric monitoring networks across the peninsula have also been destroyed, which means there is much uncertainty regarding the actual relationship between rainfall, surface runoff and infiltration. Despite very high annual rainfall it is evident that Freetown's water security should not be taken for granted.

Projecting future demands and predicting risk is fundamental for all water companies. In the case of Guma, historical feasibility assessments were completed. In 1967 the Orogu dam scheme was identified as the second major water supply option for Freetown. Further studies were undertaken in the 1980s, and a feasibility study for Orogu dam was completed in 1985 reflecting its status as a priority investment for Freetown. The civil war halted this project, however, in 2008 recommendations were made once again to safeguard the catchment area and restrict further encroachment. The Atkins study (2008) warned that: *If development continues for any length of time then the Orogu scheme may not be feasible, and massively more expensive water resource options requiring pumping over large distances will have to be developed.* Assuming that the Orogu dam is constructed, it would provide a gravity-fed supply of water, with low operation and maintenance costs, to eastern Freetown. This would enable Guma to expand its services to the east, which is a rapidly expanding area of the city.

Regretfully, repeated delays in securing and protecting the catchment area for Orugu dam has at worst made this option potentially unfeasible and at best a significantly costlier investment. The armed conflict side-tracked many of these important planning processes. However, the limited actions taken by successive government's have been a major contributing factor and external donors have not collaborated adequately to ensure financial investment and support for this critical investment. Alternative long-term solutions (other than Orugu) are expected to be either the proposed Rokel river abstraction option, with its high treatment and pumping costs, or the Bumbuna with Yiben option. This will incur higher capital costs and will likely take time to construct, but, due to the gravity flow supply, should have a significantly lower long-term operating cost. The seriousness with which governments work with water companies to advance these mega infrastructure projects is an indicator of their desire to deliver water security for their people.

The Freetown water supply network also needs infrastructure enhancements on a massive scale. Without significant capital expenditure (CapEx) it is reliant on an inadequate and aged water supply network. This propagates a vicious cycle of poor infrastructure, high non-revenue water (NRW), poor service levels, unwillingness to pay, low tariffs, low revenues, limited investments and consequently continued poor infrastructure.

Guma supported a condition assessment of its distribution network in 2018. The study identified that water pressures in the distribution network range considerably from 4.1 Bar in Lakka to less than 0.35 Bar in Freetown's Central Business District.

Water pressure in the water supply network declines rapidly from the distribution take-off points due to high pipe headloss caused by undersized pipes and/or pipe wall roughness (SMEC, 2018). The assessment also highlighted that it is unlikely that main pipeline leakage is a significant contributor to water loss as pipeline bursts only occur sporadically and are repaired quickly. However, leakage across the water supply network caused by 'spaghetti' connections and lines is estimated to be 40%.

These physical losses are estimated to be in the order of 13.3–20.4 mega litres per day (MLD), which equates to enough water to supply 140,000–225,000 people per day at 90 litres per day. However, in reality, the vast majority of consumers in Freetown receive their water supply from public standpipes or bowsers rather than household connections. This means their daily consumption is relatively low, estimated at around 25 litres per person per day (l/p/d), because water still needs to be fetched and carried back home (GVWC 2019).

The network condition assessment also estimated that eliminating illegal connections could assist in serving an additional 225,000 people in Freetown. The overwhelming contributor to leakage is the exposed 'spaghetti' service connections, which are prone to physical damage because they are not protected, often bundles of thin high-density polyethylene (HDPE) service connections which are highly prone to failure, laid on the surface or more commonly in the surface water drains, often stretching for several hundred metres with an estimated length of 990 km. Overall analysis conducted in 2018 estimates that the NRW rate, adding commercial losses to the physical losses, was estimated to be 56.5% during dry weather and 57.8% during wet weather periods (SMEC, 2018).

Unsurprisingly, Guma is still some way off its desired performance levels. Customers report just 22% satisfaction with 'quality of services' and 24% with 'reliability', district metered area (DMA) survey 2018. Those who do have a water supply service receive water on average for 4 days per week, averaging 45 hours of service over the week (less than a third of the time). Too many customer meters are not working, and the company is unable to bill according to volume consumed. Guma also has challenges in getting correct bills to customers and in making it appropriately easy and convenient for them to pay. Despite the introduction of bills' payments through mobile platforms (such as Orange Money and AfriMoney), most of the customers still pay directly to Guma offices. This in the context of the water tariff being approximately one half to one-third of comparator utilities in Africa.

Most urgently, there is not sufficient water available to supply existing customers at the level they desire, forcing Guma into a rationing regime that must suppress demand for water. Lower-income consumers are most affected by the shortage of water, particularly those residing in eastern Freetown. In the dry season (December–April) this means walking and carrying less than World Health Organisation recommended amounts of water (40 l/p/d) over significant distances in containers which may not be hygienic. This presents a major challenge to public health and a particular challenge to women and children, who generally fetch water. Developing a pro-poor service focus is therefore a key requirement for Guma, something which can only be built upon an adequate revenue-generating base of conventional customers.

3.4 APPROACHES FOLLOWED

In recent years, the policy environment in Sierra Leone has been more supportive of efforts to improve water resources management and water supply services in Freetown. Sierra Leone revised important water legislation in 2017 that provides Guma with a strengthened mandate. Sierra Leone has also embraced water and land resources management approaches by passing a new water resources law. This directly led to the formation of a new National Water Resources Management Agency (NWRMA). A new financial regulator was also established in 2014, termed the Electricity and Water Regulatory Commission (EWRC). Their role is to help Guma break the repetitive cycle of weak infrastructure, poor service levels, low tariff, low tariff revenues and low investment. However, there are ongoing constraints to implementation that include collaboration among various stakeholders and a focus on priority tasks that are important areas for strengthening resilience.

Realising these problems are chronic, Guma began a programme of step-by-step transition back in 2012. The initial intention was to motivate Guma's employees to address some of the entrenched service performance problems that persist. The process involved participatory workshops, facilitated by consultants from Adam Smith International and 2ML, so there was a sense of collective responsibility for improving the company's performance. Although the project had limited resources and duration it focussed on demonstrating that Guma had the desire to undertake necessary institutional reforms. Specifically, the following approach was implemented by the project:

- Identify, in confidence, some of the problems and poor performance ongoing within the company. This included, illegal connections, limited staff numbers, lack of resources and low levels of motivation.
- Identify realistic actions that could be achieved within the scope of the project.
- Focus, in the short term, on increasing revenue generation, including the payment of long-standing arrears.
- Sensitise employees to committing to performance targets, and set SMART (specific, measured, achievable, realistic and timebound) targets for western, central and eastern area field offices.
- Incentivise good behaviour and better performance on a monthly basis.

The workshops identified some major institutional problems, which included low motivation, low salaries, inadequate staffing levels and a sense of mistrust between senior management and Guma's employees. However, over a short period Guma were able to significantly increase their revenue generation and importantly they created the necessary awareness about the challenges faced by the company. In this respect, this was the most successful aspect of the programme because it demonstrated to donors the potential within Guma and the commitment of its staff. It raised awareness on the major water supply issues in Freetown and it encouraged external donors to re-engage with the utility, where previously there had been scepticism.

From 2014 to 2016 Sierra Leone was struggling to deal with the West African Ebola outbreak. This did not have a direct impact on water supply infrastructure, but it did mean the main water utilities and service providers were busy supplying water to hospitals and Ebola Care Facilities that had been set up across the country. Guma and other service providers also had to adopt guidelines for the management of solid and liquid wastes. Sierra Leone was the worst-affected country in West Africa and the repercussions to its people, institutions and finances went way beyond the final death toll. During this period the successful staff-incentive programme had to be suspended.

The Ebola crisis highlighted the resilience challenges in countries like Sierra Leone, but Guma were proactive in securing additional funding in the post-Ebola recovery programme. Planning issues in post emergency programmes were twofold: first, Guma needed to be able to give a clear direction as to what financial investments and support was required. Recovery programmes are designed rapidly and it cannot be taken for granted they will identify the right priority interventions. Detailed feasibility studies are important for guiding interventions, because large numbers of external consultants may be deployed with limited historical knowledge of the water situation in Sierra Leone. The second issue concerns timeframes. Recovery programmes are often over-ambitious and set unrealistic timescales for project interventions. For Guma, effective long-term development is the best form of disaster risk reduction; and sound infrastructure investments coupled with company incentives and strengthening represents the best form of disaster preparedness.

The different priorities in improving service performance and resilience exhibited by the various donors (and their consultants), along with the different time frames for project deliveries do not make it easy for a utility still in 'survival mode' to get best value and service security from donor contributions.

3.5 AQUARATING ASSESSMENT

In responding further to these long-term challenges, Guma voluntarily undertook an AquaRating assessment in 2016 – the year the Ebola crisis in Sierra Leone was declared over. This is an international rating system for water and sanitation utilities which focuses on the multi-faceted challenges they face. The rating process evaluates utility performance through indicators and management practices and defines them against an international standard. The assessments are undertaken by independent auditors accredited by AquaRating and led by the International Water Association. The primary reason for Guma using AquaRating was threefold (Guma Valley Water AquaRating Implementation, Report to GVWC & IWA/AquaRating, Eberhard, 2016):

- To establish an objective baseline of the performance of the company.
- To inform improvement action plans for Guma.
- To enable improvements to be measured over time.

Guma is the first utility in Africa to undergo this rating assessment. Perhaps unsurprisingly the overall assessment score is 10.38. This was a low score compared to the maximum score possible of 100. However, it should be noted that the standard applies to the best water utilities in the world and Guma operates in a volatile and challenging context with many known performance problems. A low score was therefore anticipated by the assessors and Guma staff.

It was also recognised that the derived score was in-fact less important than the learning that can be gained and the subsequent improvements made as a result of the assessment. It can also be noted that a lower base score provides for greater room for improvement. The assessment also suggested that Guma was being transparent and opening itself up for wider scrutiny, which is of far more interest to investors than assumed information.

Scores by assessment area are shown in Figure 3.2. It is notable that the scores are variable across the areas with the highest score given for corporate governance (65) and second highest for business management efficiency (32). The other six areas scored below 19 with very low scores (below 10) for access, operating efficiency,

Figure **3.2** GVWC AquaRating assessment scores 2016. *Source*: Eberhard, R. (2016) Guma Valley Water Company Aquarating Assessment, Available from Guma Valley Water Company, Freetown Sierra Leone.

and environmental sustainability, and a score of 13 for service quality, 15 for financial sustainability and 18 for investment planning.

To put the AquaRating assessment into context it is helpful to highlight two of the very real challenges Guma face. For example, as Freetown's population has grown to 1.5 million, the number of household connections has stagnated to around 22,000 with just 200 actual volumetric (metered) customers. To address this problem, Guma introduced a sophisticated system of negotiated fixed rate tariffs in order to help generate much needed revenue. However, externally implemented programmes have, at times, proposed the widespread introduction of water meters, yet this could actually serve to undermine Guma's revenue when available water supply is so limited. A second example is the ability to increase revenue by serving customers in the Central Business District (CBD) of Freetown. For some time now Guma's own headquarters office, located in central Freetown, has been reliant on bowser water delivery. This is an indication of the company's struggles to prioritise water supply to the best potential revenue stream (business customers) in the CBD. This would also serve to boost the morale of Guma staff members by improving the functionality of company toilet facilities.

Guma's staff are often frustrated with the current circumstances. They are trying to do their best for customers and consumers but are often restricted by the system in which they work. Salaries have been below national and international comparators and they lack adequate equipment, logistics and finances to perform essential utility functions routinely. In the recent past they have also been criticised by the national media and government. Some of this criticism may be justified but not entirely. The major challenge Guma continues to face is the shortage of water to supply customers as growing demand continues to outstrip available supply. Guma has advocated repeatedly for catchment areas to be protected and has requested direct support from mandated Ministries, Departments and Agencies responsible for law enforcement, land management and preventing environmental degradation. Without increased support to safeguard raw water resources it is not realistic to think Guma can work independently to be a resilient utility, because a lack of water, coupled with limited political will to support the water sector directly affect all aspects of their operations.

3.6 INSTITUTIONAL REFORM

Institutional reform is primarily concerned with transitions. The change from one (inferior) situation to a much better one. The intention for Guma is to demonstrate a vision and real commitment to deliver water for all so it can attract ongoing technical and financial assistance. The process involves Guma doing *what they can with what they have,* with each department and unit developing detailed workplans, focus areas and key performance indicators. These timebound plans are detailed in Guma Valley Water Company Strategic Performance Improvement Plan (2019–2023).

GVWC is committed to delivering the internal reforms required to improve services for customers and to provide assurances to all stakeholders that the company is committed to supporting Sierra Leone's national targets and the sustainable development goals. However, GVWC cannot meet these longer-term goals without government assistance and external support to finance and develop new water sources under larger capital investment programmes.

In the short-term, delivery of this plan along with many other planned improvements is subject to Guma receiving approval to increase water tariffs by the EWRC. While the company benefits from the low operating cost of the Guma dam, which enables a gravity-fed system, Guma has only been able to operate on a very limited tariff by not undertaking critical investment in capital maintenance to the supply network. 'Spaghetti connections' are a visible sign of the Company's inability to install mains pipes in the city's streets. All of the necessary reforms need to be supported with improved customer engagement and quality data on existing situation to ensure buy-in and support and to explain the reason for the necessary increase in cost. The first tariff revision submission to the new economic regulator was requested to be postponed for over a year.

One way Guma has tried to address these issues is to set out its vision and ambition in its 2019 Business and Investment Plan. This plan sets out Guma's starting point and details the challenges they currently face, their future vision and ambition, and how they intend to progress towards these timebound targets. The plan is ambitious, but it sets out the utilities vision for transformation. It is structured around five main corporate objectives (COs):

(1) To improve core utility operations and customer care, including to the poor.
(2) To empower staff: strengthening HR capacity and incentivising company productivity.
(3) To achieve financial sustainability and increase investment planning.
(4) To enhance water resources and network infrastructure; expanding coverage.
(5) To develop professional corporate support services and information communications technology.

3.7 INFRASTRUCTURE

Guma's improved performance is directly dependant on repairing, rehabilitating and replacing critical infrastructure. Immediate problems to address include: massively increasing new raw water storage capacity, replacing large sections of the transmission and distribution network, utilising the full capacity of existing infrastructure, increasing water pressure in select areas, repairing pumping stations, and improving access for customers and consumers. At present, water supply is rationed year round to try to maximise efficiency and equity. With a population growth rate around 3% per annum investment and infrastructure work is pressing to avoid growing numbers of unserved.

The lack of investment in water infrastructure has created the potential for significant water contamination, particularly in the wet season. According to data collected during the network condition assessment, there are 18.1 km of submerged distribution pipes, the majority of which are located in storm drains and ditches on the side of the road. Both pipes in submerged ditches and leaking spaghetti connections create potential contamination points due to contaminated water being drawn into the mains when pipes empty due to rationing cuts. This is a significant issue given high rates of open defecation in Freetown and the inadequate sanitation systems for routinely removing, treating and disposing of solid and liquid waste in a safe and responsible manner.

3.8 ACHIEVEMENTS SO FAR

Building resilience of water supply services in Freetown will take considerable time. It is therefore concerned with transitions. Not all of these transitions have been completed yet, but the process improvements being undertaken by Guma point to a more resilient water utility.

The first transition, which is perhaps the most visible, is the replacement of aged and crumbling infrastructure. Guma has been working closely with IMC Worldwide, BAM Nuttall and Mott MacDonald (and others) to rehabilitate its Water Treatment Plant and replace large parts of the water supply network. The Freetown water supply rehabilitation project, which aims to improve water quality, reduce leakage and extend water supply to eastern Freetown, is the first of many much needed infrastructure projects.

Guma is also partnering with Millennium Challenge Corporation to improve capacity and to develop core systems and processes to improve services. This includes establishing 'District Metering Areas' (DMAs) to reduce NRW in two pilot areas in Freetown, with the broader goal of replicating the process across GVWC's entire network. This should be completed over the coming decade, anticipating distribution of additional treated water from the planned and much needed new water source/s. Also, African Development Bank (AfDB) and partners are sponsoring the Freetown WASH and Aquatic Environment Revamping programme. This work aims to deliver the rehabilitation of five malfunctioning raw water intakes and the development of 10 new water intakes, along with 30 km of raw water transmission mains, to provide approximately 51 MLD additional potable water supply, distributed through the rehabilitation of 100 km of distribution mains and the expansion of the distribution network by 150 km, delivering through 56,000 new metered household water connections. The approximately $192 m budget for this work (which also includes watershed protection and sanitation elements, spread over 48 months) has been incorporated in the 2020–2025 Business & Investment Plan, assuming a four-year implementation period. The project is currently in its inception phase.

In advance of future major water resource investments, GVWC has also planned to invest, with Government support, approximately Le 100 billion in small-scale water source development and enhancements wherever possible on the Peninsular (Kissy, Dwarzak & New England Communities, Kossoh Town, Hastings Town & Jui, Boama, Angola Town & Philip Street in Wellington; approximately 5 MLD total) and other hard-to-reach communities within Freetown. However, that addition, along with water conserved from fixing leaking pipes and connections, will not be sufficient to meet the immediate shortfall of over 100 MLD, let alone for the future overall demand as the newly mandated (GVWC Water Act 2017) Western Area population grows. Guma is also planning in advance for the next major investment in water resources. This is being achieved with the support of the African Development Bank (AfDB) who are funding the 'Freetown Water Master Plan & Sanitation Master Plan and Investment Studies'.

The oft-repeated saying is 'software without hardware goes nowhere' which implies ongoing support to Guma should combine both capital investment and utility reform to maximise impact. The second transition therefore has been to improve knowledge so management decisions and utility performance improves. Extensive work has been undertaken to help Guma better understand the location and condition of its water supply network. Infrastructure has been mapped extensively and a detailed condition assessment completed to inform Guma about the status of its critical infrastructure assets. Flow monitors have been installed to help Guma understand 'where the water is going' and areas of 'low flows' in order to improve water distribution. Based on this recorded flow data, Guma has been able to revise its rationing regime and inform people when their localities will receive water.

If there is ignorance concerning the status of assets and water supply then there can be little hope of managing services well and making informed decisions. But, the reality is that multiple processes also need to be improved to establish asset management systems, improve customer engagement and use survey and real-time data for better decision-making. In the past, Guma's core utility operations have not performed well. A lack of data collection, analysis and follow-up action, limited systematic process and a lack of resources meant it was not easy for Guma to improve its operational performance. But times change and embedded technical support programmes and new technologies have allowed Guma to begin to address some of these entrenched problems. A third transition, ongoing within Guma, is to improve its management processes in a systematic manner. There has been significant progress in embedding data-driven management processes that will enable GVWC to achieve improvements in its operations. Data and information have been emphasised as a core driver of utility performance. For example, an initial assessment found that many existing information systems and databases were underutilised and there has been a concerted effort to improve the way these data systems are utilised and applied. Guma's Customer Information System (EDAMS), for example, is a robust system that already has a lot of valuable data that can provide the utility with critical insights on how the water 'inventory' is converted into revenue. The GVWC Operations Report relies on data from EDAMS to develop the water balance and to provide analysis on its DMAs. Used effectively EDAMS also provides critical inputs for non-revenue water management and with technical support from Adam Smith International, Guma has developed a process for dealing with customer queries in order to generate reports for senior management and strengthen the companies standard operating procedures (SOPs). Further, critical information has also been collated on the status of water supply network assets, including the generation of visual GIS maps and a condition assessment of all pipes, reservoirs, valves, and other critical operational assets.

Due to the extensive number of illegal connections and extensive open defecation water quality is also problematic across Guma's water supply network. New water quality management process have been introduced. This required a process of: first embedding new SOPs to test water regularly at key locations across the network; second analysing the data and ensuring corrective follow-up action to improve water quality. This has required substantial resource increases in equipment, transport, and laboratory facilities, including on-site portable testing. Guma has worked intensively to ramp up testing and field operators are now able to apply automated analytical tools in Excel to help GVWC quickly produce summary reports from the raw data. This has already had a positive impact on the ground, where Guma has invested in cleaning and safeguarding service reservoirs at Tower Hill.

An important way in which Guma have tried to drive performance is through the introduction of an improved results based financing (RBF) programme. This programme or intervention has primarily focussed on addressing revenue collections and technical losses and has been one positive way to incentivise better company performance and improve staff morale. However, RBF programmes can have a positive but temporary impact if they do not focus on improved processes rather than just outcomes. This is because outcomes cannot be sustained unless internal operational, commercial and financial processes are improved. An important way the RBF programme has been pursued is through linking it to a Performance Incentive Sharing Plan (PISP). This has been achieved by engaging with area managers (western, central and eastern) so they can play a more meaningful leadership role in helping the utility to meet its objectives. This has been achieved through collaborative planning, implementation, data collection, analysis and distribution of incentive payments, based on performance targets and results. While the PISP programme will not resolve all problems it is an important first step in problem solving, improving processes and motivating staff. It also challenges improvements in management and oversight in order to build trust between frontline staff and senior management. Examples of initiatives being taken include: improving response times to reported leaks, installing new sub mains, improving water quality, handling customer complaints efficiently, and registering new customers.

There has also been a transformation in Guma concerning customer engagement, particularly since the appointment of a new managing director in July 2018. This has focussed on prioritising customer engagement but crucially also with corresponding changes in Guma's own accountability and behaviour. This is evidenced by work in the two DMAs to manage leakage, change Guma's behaviour and to develop better customer engagement strategies. The GVWC Stakeholder Engagement Plan now embeds the concept of 'Key Influencers', for example, which emerged as a very useful strategy in the pilot DMAs. Part of this transformation in customer engagement also includes outreach and using media channels to engage with the public more widely.

The introduction of two DMA demonstration has provided multiple benefits. They have created a practical learning-by-doing environment that has been a critical part of Guma's institutional strengthening strategy. The process of planning, developing, and implementing pilot DMAs required an extensive effort to meaningfully understand services and challenges on the ground. The process required digging deep into data that is stored in EDAMS and provided the team with an understanding of how to use data to drive service improvements on the ground. An outcome of the DMA work has been the ability to categorise customer accounts into: (1) accounts billed and paid, (2) accounts billed but not paid, and (3) accounts not billed – usually because water was not supplied. This has allowed the Guma's area teams to improve the correlation between service levels on the ground and customer payments.

The DMA work and analysis also demonstrated that pro-poor services cannot be treated as a niche. Over the past few years poorer communities have always been served by water bowsers, which are often discontinued due to high operating costs and breakdowns. This is not a sustainable model. It does not help poor people and it does not help GVWC. In response the primary objective of support to CWS now has been to reorient the bowser unit to a strategic unit that supports area teams in extending services to ALL residents of Freetown. The efforts around embedding this internal change have been driven by work under a social behaviour change and communication programme to influence behaviour change within the utility. The CWS unit now works with area teams to identify supply gaps and then help the teams assess options for extending services. Where possible, area teams must provide household or yard connections and where this is not possible, there are a menu of options that include meter banks and water kiosks.

A fourth transition has been to focus attention on Guma's own human resource systems in order to improve utility performance. Staff motivation and capacity continue to be major challenges that GVWC need to address. There are four aspects in particular that Guma has focussed on. The first has been to introduce a performance incentive sharing programme (PSIP), which focussed on improving revenue generation and reducing technical losses across the water supply network. The core intention of staff incentivisation is to drive performance improvements and the concept builds on earlier ideas of empowering GVWC staff and a results-based framework that incentivises good performance on a monthly basis. Area managers and area teams, who interface with customers on a daily basis, in particular need to be empowered to play a more meaningful leadership role in helping the utility meet its objectives. While the PSIP will not resolve all the issues that prevent this from happening, it is a useful starting point. The second aspect has been to improve HR processes and disciplinary mechanisms. Detailed HR manuals have been developed and line managers supported to ensure higher professional standards, where previously working arrangements and the influence of trade unions were considered unworkable. GVWC staff now have an HR handbook that informs employees of their contracts, benefits, disciplinary and grievance mechanisms. The third aspect has focussed on reducing and eliminating gender-based violence including policies and SOPs related to sexual harassment, bullying and discrimination to prevent incidents for both GVWC staff and its customers. These important measures are all part of a wider initiative to substantially address the way the utility operates. The fourth, relates to the introduction of ongoing continued professional development, which includes practical embedded training, study tours to other utilities and a series of thematic Master classes for managers. GVWC has also formed stronger links with the Sierra Leone Institute of Engineers so ongoing utility training is recognised and accredited.

The fifth transition has placed the spotlight on financial sustainability and investment planning. If Guma is not financially sustainable then there will be little prospect of building resilience for future shocks and threats. A number of initiatives are ongoing that include: first, improving financial management systems within Guma. This has included internal restructuring within Guma to strengthen the Finance and Administration Department, introducing new financial management systems (such as SAGE) and improving financial management capability. The Finance & Administration Department are expected to provide strategic and analytical support to provide better financial analysis, cost–benefit analysis and tracking staff productivity. The second is about sound investment planning. Priority-based budgeting has been introduced so scarce resources are targeted to maximise impact, leveraging new sources of financial investment; and moving towards a new cost-reflective tariff area all efforts to generate revenue so service improvements are undertaken. Guma's proposed tariff increase will maintain the existing lowest domestic tariff to ensure lower-income customers with household connections can access water at an affordable rate. It is anticipated that this rate will eventually be reflected in the charges for water from stand-posts and kiosks. In future the company needs to generate sufficient revenue to ensure that existing systems can sustainably deliver to customers at its renewed and rehabilitated state with ongoing regular maintenance.

3.9 OUTCOMES

Guma are now recognised as being serious minded about improving water supply services in Freetown. Government also realise that water resources management and water supply in Freetown is at a critical stage. Across the city, people openly talk about annual water shortages, frustrations with service levels and inequity across the city. Businesses have spoken about willingness to pay for a more reliable service. Now people start to see action on the ground. One clear indication is the rehabilitation of the main treatment works at Guma dam and the rehabilitation of transmission and distribution lines. The separation of commercial bowser operations from the community water service delivery has increased efficiency in the delivery of water to community tanks and private individuals. In early 2021, Guma received 13 new water bowsers and now have 18 functioning bowsers to deliver water to communities, especially those without piped connections. A number of projects are introducing new water kiosks and a new rationing regime has been designed by Guma so people know when water can be expected in their local area. Moreover, infrastructure feasibility studies are planned to determine the most viable and cost-effective long-term water supply options.

The formation of the two regulators, NWRMA and EWRC, is also an indication of government's desire to ensure sound stewardship of water resources and a fair deal for consumers. Guma now interacts with both regulators with the support of the Ministry of Water Resources. This provides an opportunity to coordinate activities and ensure meaningful progress on some key issues. These multi-stakeholder platforms need to become more prominent in the future and would benefit from ongoing engagement with Freetown City Council, which is led dynamically.

As part of the ongoing Freetown water supply rehabilitation work, multi-stakeholder platforms and local consultations have also been held with customers and local residents. Working collaboratively is viewed as a way to resolve water issues at a local level. This includes catchment encroachment, illegal land occupations and vandalism to the main water supply network.

There is still a considerable amount of work to do over many decades. However, this early progress has attracted growing donor interest and many are committing to further programmes of technical assistance. These programmes also need to empower Guma staff so they have better working conditions and are incentivised to participate. Incentive-based performance contracts have already been introduced within Guma at the organisational and institutional level, and should be considered routine to drive the desired change. As an example a staff loan scheme has also been introduced with 10.5% interest rate payable within 18 months to motivate staff.

3.10 KEY LESSONS
3.10.1 Ownership of the change process

When countries are susceptible to repeat shocks and crisis' it is vital that all water sector stakeholders (such as government, water companies, external donors and INGOs) collaborate meaningfully so they recognise which actions will deliver maximum improvement. The challenge that Guma has often faced is that many development and humanitarian support programmes are planned remotely, which makes it difficult for Guma to communicate their exact requirements. It is only when the most pressing actions are identified that utilities can build their capability in a step-by-step manner. There are also problems around quick fix programmes trying to address long term, chronic issues. Often these short-term interventions fail to deliver the necessary impact and staff are sometimes resistant against changes that transform their usual ways of operations.

3.10.2 Political commitment

Government commitment to supporting utility reform is crucial. Utilities require support to protect their catchment areas to secure raw water resources. Support from the executive level of government is also required to advance discussions and planning for mega infrastructure investments. High-level consultations are required so pre-feasibility and feasibility studies can be undertaken on time and investment finance secured. There have been successive studies in Sierra Leone reconfirming the importance of the Orugu dam catchment site. Guma senior management has for many years drawn attention to this critical issue, but successive governments have struggled to secure the catchment area and no detailed designs have been undertaken following the earlier feasibility work. Urban water supply should be strengthened through supporting water utilities rather than politicising the provision of water to the population on a free of charge basis.

3.10.3 Infrastructure performance

Guma's performance is heavily dependent on the availability of adequate water resources and the state of their physical infrastructure (termed assets). Systems need to be put in place to help utilities oversee the planning, design and construction of new infrastructure, such as reservoirs, treatment plants, transmission and distribution systems.

With investments required in the hundreds of millions of dollars, government and donors are understandably reluctant to invest without seeing demonstrable institutional change. But with such capex programmes necessarily taking a decade to implement, at a minimum, stakeholders have to commit in advance.

One approach Guma has adopted is to divide its water supply network into a series of DMAs. This enables the utility to repair and upgrade the network in a step-by-step manner to improve service levels and address leakage. This work will also generate information on the utilities non-revenue water situation. However, the utilities desire to improve service levels and generate revenue within DMAs, should also be tempered with a focus on serving the urban poor and low-income communities. In Freetown, these community groups will likely remain dependent on standpipes and tapstands for the foreseeable future and the charge per bucket should be equivalent to an appropriate subsidised lifeline tariff.

If cities and towns experience high levels of social inequality there will be little prospect of a resilient society. Vulnerable communities will likely live in densely populated and unhygienic slums that are prone to flooding or environmental pollution. Utilities and the economic regulator should collaborate to develop policies, tariffs and innovative solutions that are relevant for households and communities that are in the lower wealth quintiles. Utilities often lack organisational capacity in gender, equity and social inclusion (GESI) matters and specific units may need to be established that focus on serving vulnerable groups in a fair and equitable manner.

3.11 INSTITUTIONAL REFORMS TAKE TIME

There is a critical need to move away from concentrated large-scale institutional reform processes that exceed a utilities receptive capacity. It must be recognised that utility staff have their own day-to-day roles and responsibilities to perform and it becomes a major challenge when multiple donors embed technical support programmes simultaneously. Donors should also avoid investing their efforts in producing multiple technical, social and financial reports that will probably serve a limited purpose within the utility. This inadvertently creates a problem whereby participation and trust building becomes difficult because technical support programmes are unrealistic in their timeframes and do not recognise the need for long-term support.

3.12 COMMUNICATION WITH CUSTOMERS

Water utilities are accountable to their customers first and foremost. Customers have often struggled with low service levels and have not had a direct line of communication with Guma to demand better services. Previously there were no senior positions within Guma whose primary role is customer service and communication. Even though it is not an easy task, a customer charter could have enshrined Guma's commitment to the customers and people in Freetown at an earlier stage. Potentially, this could have been crucial in leveraging greater support from government. However, Guma's weekly radio programmes have created a platform for customers and consumers to express their grievances and concerns about the company's services.

3.13 SUMMARY

To summarise, the case from Freetown demonstrates that water companies in fragile states are starting from a low-base. All aspects of their business need substantial long-term technical, institutional and commercial support. Building resilience will take a significant period of time (decades rather than years) and it will require investments in many essential areas, such as water resources, infrastructure, institutions and finance. The key issue is arguably to support the utilities processes to develop a forward looking business and investment plan, which can be used to articulate and communicate the utilities own aspirations. It should be used to generate greater political and donor support.

In a long-term development setting it is vital that the utility can communicate what are the single most important areas for improving performance levels and resilience. As offers of financial and technical assistance are made, both utilities and government need to be more coordinated and adept at communicating what assistance is actually required and when. This is important because individual support programmes cannot deal with the entirety of infrastructure, institutional and commercial problems. Therefore, part solutions must be well coordinated and phased in a logical order. Guma's experiences show that historically opportunities or critical junctures to ensure long-term water security have been missed, which will now result in more costly interventions and operations. The failure to secure and protect Orugu catchment area and build a second major dam and reservoir for Freetown is the obvious example. This inadvertently hinders the prospect of long-term resilience.

REFERENCES

Eberhard R. (2016). Guma Valley Water Company Aquarating Assessment, Available from Guma Valley Water Company, Freetown Sierra Leone.

Guma Valley Water Company (2019). Strategic performance improvement plan 2019–2023. Available from Guma Valley Water Company, Freetown, Sierra Leone.

Guma Valley Water Company Act (2017). Guma Valley Water Company, Freetown, Sierra Leone; available at: http://extwprlegs1.fao.org/docs/pdf/sie176436.pdf.

Ledger D. C. (1975). The Water Balance of an Exceptionally Wet Catchment Area in West Africa. Available at https://www.sciencedirect.com/science/article/abs/pii/002216 9475900815 (accessed 07 April 2021).

Ministry of Water Resources (2015). Strategy for Water Security Planning, Volume 1 of a three-volume set, Sierra Leone.

SMEC (2018). Guma Valley Water Company Water Supply Network Condition Assessment.

Statistics Sierra Leone (2017). Sierra Leone Multiple Indicator Cluster Survey, survey findings report, available at: https://www.statistics.sl/images/StatisticsSL/ Documents/sierra_leone_mics6_2017_report.pdf.

WS Atkins with 3BMD and Oxfam (2008). Strategic Water Supply and Sanitation Framework Parts 1, 2 & 3. Available at: https://www.coursehero.com/file/902067 80/Atkins-Guma-watersupply-and-sanitation-framework-Freetown-1pdf/.

Chapter 4

Mobilising the public to reduce household water use in Essex and Suffolk Water

Fatima O. Ajia[1], Tim Wagstaff[2] and Liz Sharp[1]
[1]*The University of Sheffield, Sheffield, UK*
[2]*Essex and Suffolk Water, UK*

ABSTRACT

The south-eastern region of the UK is facing water scarcity due to population growth and insufficient rainfall to meet household water demand. One of the regulatory requirements for water utilities is customer engagement to increase water efficiency. This chapter aims to identify key barriers to delivering engagement activities promoting household water efficiency and opportunities for improving practices in Essex & Suffolk Water (ESW) – a UK water utility operating in areas of serious water stress. A reflection is made on the water utility's Every Drop Counts (EDC) home visit campaign, an annual household water efficiency initiative, with particular focus on insights from its face-to-face delivery during Asset Management Plan 6 (AMP6, 2015–2020). The pilot of the EDC campaign's virtual initiative comprising of 66 virtual home visits is examined, with focus on drawing out lessons learned as Asset Management Plan 7 (AMP7, 2020–2025) begins during the coronavirus disease 2019 (COVID-19) pandemic.

Whilst the virtual home visit campaign was found to reach a broader customer base, save financial and environmental costs, and address the season and place constraints typically posed by the face-to-face campaign, fewer water saving devices were installed per property (4.4) compared to the face-to-face campaign (6.4), and calculating measured water savings was impossible due to customers failing to take water meter readings independently during the COVID-19 lockdown. Face-to-face home visits should therefore not mean an end to virtual home visits and vice versa, but rather serve as a twin-track strategy for delivering the campaign.

Key strategies that emerged as improving face-to-face home visits in ESW include increasing the use of customer insight; varying the frame for water efficiency communications; improving the face-to-face engagement strategy; enhancing knowledge training; and creating feedback mechanisms between water efficiency managers and plumbers on the frontline. To better maximise virtual home visits, it is recommended that the behavioural change aspect of water efficiency education is delivered as a key and complementary aspect of appointments, and customers are better supported to self-install a wider range of water saving devices.

This chapter bridges the gap between water management theory and practice by providing a better understanding of how practitioners are putting concepts into action on the ground and by so doing, contributes to building a learning culture in the global water sector.

Keywords: water demand, water scarcity, water efficiency engagement.

4.1 INTRODUCTION

Water scarcity is an issue set within a socio-environmental context in the UK. According to the Office of National Statistics (ONS) (2018), the UK population is at an all-time high due to births and immigration outweighing deaths and emigration. The ONS projects that the UK population will increase to 73 million people by 2041, a 46% growth in seven decades. In response to population growth, total household water use is expected to rise. The densely populated south-eastern region of England already has an average annual rainfall which is around 500–600 mm less than Sudan, or Perth, Western Australia. There is therefore insufficient fresh water available to meet people's future needs. Adding complexity to this problem is climate change, which is expected to further reduce supply and may also contribute to increasing demand (Adaptation Sub-Committee, 2016).

Since the privatisation of the UK water industry in 1989, water utilities undergo a 'Price Review' (PR) every five years, wherein the economic regulator, Ofwat, assesses the companies' business plans to set a price limit for water. Following the completion of a PR, the five-year period within which the water utilities deliver their plans is known as the Asset Management Plan (AMP) cycle. The UK water sector completed its seventh cycle of PR in 2019 (PR19) and the delivery of AMP7 is now underway.

The significance of the PR in this discussion is that it defines the period for water resources planning (Hamling *et al.*, 2018). Water resources planning is key for water efficiency engagement because it is the process through which the UK environmental regulator, the Environment Agency, assesses water utilities' Water Resources Management Plans (WRMPs) to balance water demand and supply. Companies achieve this balance using a range of interventions including water efficiency engagement. For their customer engagement activities during the previous PR in 2014 (PR14), which is discussed below, water utilities were guided by Ofwat's (2011b) policy (which gives guidelines about who to engage, what to engage about, when to engage, and how to engage). And their customer engagement activities for long-term resilience during AMP6 drew guidance from this policy.

Following PR14, the UK water sector regulator, Ofwat, has continued to task water utilities with strengthening their resilience in the face of threats to water resources. An aspect of this requirement is the promotion of water efficiency via 'education' initiatives intended to encourage the public to reduce their water usage. Typically, these water efficiency education activities are expected to increase the public's awareness and knowledge about water and motivate them to change the way they use water with the hope that a reduction in per capita consumption (water usage) is achieved. However, many water efficiency education initiatives fail to fulfil their demand management potential. Difficulties arise because the attention is entirely focussed on the public's responsibility to act for change. The role of water utilities in impacting on the public's water usage and efficiency is not often at the fore of project-planning. There is therefore the potential to increase the effectiveness of utilities' water efficiency education practices.

Essex & Suffolk Water (ESW) is building resilience to water scarcity through water efficiency education on the frontline. However, there is a recognition that details about education strategies and techniques used by water utilities to support household water efficiency are not robustly shared across the water sector, and when shared, do not highlight challenges faced and lessons learned.

Guided by the message-audience-channel (MAC) heuristic for understanding water efficiency engagement, the purpose of this chapter is to present and reflect on ESW's water efficiency education activities, particularly in relation to the Every Drop Counts (EDC) home visit campaign. The MAC heuristic is a framework for examining water efficiency engagement practices in the context of who is engaged, what the engagement is about, and how the engagement is delivered (Ajia, 2021). The water utility's education practices undertaken during AMP6 (2015–2020) are examined. Barriers to effective water efficiency education on the part of the water utility are identified. Practice improvements that the utility implemented during the AMP6 cycle are shared. Further, the challenges to the EDC home visit campaign brought about by the coronavirus disease 2019 (COVID-19) pandemic are identified, and innovations emerging as water efficiency education is delivered in the current Asset Management Period 7 (AMP7, 2020 to 2025) are expanded on.

The next section contextualises the problem of increasing household water demand in the UK. This is followed by a review of the main water efficiency engagement approaches taken across the water industry in the third section. The fourth section gives a collective account of the EDC home visit campaigns delivered during AMP6 (2015–2020) with particular focus on engagement experiences on the frontline, challenges faced, and learning points. The penultimate section covers how the water utility is adapting water efficiency education considering the COVID-19 pandemic, culminating in a conclusion to the chapter in Section 4.7.

4.2 THE PROBLEM OF INCREASING HOUSEHOLD WATER USE IN THE UK

The Environment Agency and Natural Resources Wales (2013: 2) classified certain areas in the South East of England as experiencing 'serious water stress' because their 'current demand for water is a high proportion of the current rainfall available, and/or because the future household demand for water is likely to be a high proportion of the rainfall available'.

And although eight additional water utilities have been classified as seriously water-stressed for metering purposes, the focus here is solely on classification due to insufficient rainfall to meet water demand (The Environment Agency, 2021). In addition to causal factors such as climate change and population growth, securing the UK's future water supply is challenging due to planning, legal and public challenges that have prevented the building of new reservoirs for decades, low levels of water transfer from areas having surplus water to those in deficit, the non-operation of the major desalination plant in London due to cost implications, wastage of public water supply via leaky infrastructures, and the high levels of water usage in households.

In this chapter, of concern is the centrality of household water use to the issue of water stress in the UK. Evidence from the Department for Environment Food and Rural Affairs (DEFRA) (2018) suggests that while usage in the UK's industrial sector has continued to decrease in recent years, reductions in household water use in recent years have flatlined or even increased. Specifically, in England, the consumption level of an average person (143 litres of water per day) is higher than the UK Government's aspirational target of 130 litres of water per person per day (l/p/d) by 2030 and is also 85 l/p/d higher than the average usage recorded in the 1960s (Lawson *et al.*, 2018). England's average per capita consumption between 2015 and 2020 stalled between 139 and 142 l/p/d. Of greater concern is that in the South East of England, the case study utility, ESW, has seen a steady annual increase in per capita consumption of 1 l/p/d from 2015/2016 up until the outbreak of the COVID-19 pandemic. Since then, ESW has seen an approximate 10% increase in per capita consumption.

Unmanaged household water demand poses a threat to the UK's water security and has been described by authors such as Browne *et al.* (2013) as one of the most significant concerns for UK water utilities. The necessity to act by involving the public in water efficiency becomes more apparent when the occurrence of drought events (such as seen in 1995–1996, 2006, and 2010–2012) causes the Government and water utilities to call upon the public to urgently reduce its usage. For example, due to dry periods and high levels of water demand, the past two decades have seen increased public appeals by ESW urging people to use water less and differently.

The need to address the problem of increasing household water demand is being driven by regulation. While the water resources planning conducted by UK water utilities for PR09 (AMP5) showed more reliance on supply-side measures than demand-side measures, following PR09, the UK Government advocated increased customer engagement to encourage water efficiency in water utilities (Ofwat, 2011a). A decade ago, Owen *et al.* (2009) conducted research for DEFRA to examine the public's understanding of sustainable water use in the home. The authors found that people were not fully aware of the water situation in the UK and lacked the knowledge and motivation to reduce their usage. Their report attributed high per capita consumption partly to the low value people put on water which means that the public uses the resource without giving much thought to their usage. It is in response to such findings that the Environment Agency and Natural Resources Wales (2013: 4) recommended that in all water utilities, 'there should be some activity to ensure that water is used more efficiently and effectively'. However, a review of seriously water-stressed utilities' plans for water efficiency education in PR14 conducted by Ajia (2018) indicated that the newness of public engagement to reduce household water demand means that the practice in the UK water sector is still in the developmental stage. Post-PR14, water efficiency education in water utilities is being increasingly promoted by regulatory and academic stakeholders in the water industry as a measure to ensure a resilient water system.

Average per capita consumption in ESW in 2019/2020 was 155 l/p/d. The water utility continues to seek sustainable ways to secure future water supplies, and people that save water are a key focus. Amongst other demand-side interventions such as leakage and wastage reduction, using water efficiency education to address customers' water-related behaviours is one way to reduce demand and meet the regulatory expectation to increase household water savings.

The next section reviews more broadly, the demand-side resilience strategies that appear to be prioritised in the water management literature.

4.3 CURRENT WATER EFFICIENCY ENGAGEMENT APPROACHES IN THE UK WATER INDUSTRY

Water efficiency engagement refers to water utilities' actions and interactions to control the public's use of water and motivate behaviours and usage practices that can result in a reduction in per capita consumption (Ajia, 2021). In the UK, water utilities are adopting a twin-track approach (supply-side and demand-side) to balance water demand and supply and to ensure a resilient water system. Such interventions include seeking new water sources, reducing leakage, increasing meter penetration, and promoting water efficiency. Although water efficiency education is the focus of this chapter, it is useful to reflect on the ways through which water utilities seek to increase household water efficiency in the demand management landscape.

Currently, water utilities use water saving devices, education, and the smart water meter to achieve water efficiency. The reduction in water usage achieved using water saving devices or the smart water meter has been referred to as techno-efficiency by Browne *et al.* (2019) while water savings derived from education that influences the behaviours of water users is understood to be edu-efficiency (Ajia, 2021). Ajia categorises water efficiency engagement into four approaches based on the interventions used to seek water efficiency: technical approach; educational approach; combined approach; and sociotechnical approach. These will now be discussed within the UK context, in turn, in the subsequent subsections.

4.3.1 Technical water efficiency engagement

Technical water efficiency engagement is typified by metering or retrofitting household water systems with water saving devices to reduce per capita consumption.

The water efficiency literature suggests that although the initial aim for metering was to simplify billing, the smart water meter evidently 'motivates' people to reduce wastage and pay attention to how they use water. The UK Government considers metering to be an effective measure for reducing water demand and evidence shows that usage is lowest in the most metered areas (Parliamentary Office of Science & Technology, 2012). This is because installing a smart water meter in a household implies billing accuracy and this has a psychological effect on people. It is as though the smart meter 'speaks' to the public in a language it understands when authors such as Orr *et al.* (2018) assert that metering makes people 'change' their usage pattern so as not to pay more than they desire.

Another popular form of technical water efficiency engagement across all UK water utilities today is the retrofitting of household water systems with water saving devices to reduce the amount of water needed for domestic usage activities. Retrofitting is claimed to be capable of delivering up to 50% water savings (Dworak *et al.*, 2007), although some experimental studies particularly in the UK only reported 4–6% reduction in per capita consumption following the installation of water saving devices in households (Smith & Shouler, 2001; Keeting & Styles, 2004). Reasons for the low yield of water savings from retrofitting can be found in critiques of the water efficiency engagement approach. For instance, Knamiller *et al.* (2006) brought the long-term sustainability of water savings realised through retrofitting in ESW into question when the authors found that water usage returned to historic levels because people lacked the awareness about their retrofitted devices. Further, Browne *et al.* (2019) criticised retrofitting for reproducing individualistic paradigms of behaviour change.

Collectively, metering and retrofitting interventions to reduce per capita consumption, if done in isolation, present the risk of reducing or completely excluding the interaction between water utilities and the public as such interventions can 'mute' both parties in the engagement process. The challenge is that when people's behaviours and values relating to water use are not addressed, their usage may not change if circumstances were different. For example, the usage of a person living in a metered but rented bills-inclusive property may not decrease because usage is not personally paid for. Likewise, whilst a person's water usage could reduce following the installation of a low-flow shower head at home, their usage attributed to showering may differ when in an environment with inefficient water systems. This raises a pertinent question about what is truly central to the achievement of water efficiency – technology or people, or both?

4.3.2 Educational water efficiency engagement

The consensus amongst social scientists is that addressing people's conscious behaviours is equally as (if not more) important as using technical measures to achieve household water efficiency. This notion has popularised interventions that seek to 'educate' the public about water to motivate them to reduce their usage.

Behavioural change education feeds into DEFRA's wider policy agenda to enhance efficient and sustainable water use by addressing fundamental psychological factors that influence water usage. The educational water efficiency engagement approach is typified by Behaviour Influencing Tactics such as information sharing, awareness building, persuasion and so on, with these strategies embedded in messages communicated to people to cause them to take a desired line of action (see Koop *et al.*, 2019). Unlike two decades ago when drought events in the UK were met with light-touch campaigns such as the distribution of water efficiency leaflets to the public, Waterwise (2013) reported that response to the recent 2012 drought event in the country and other spates of low rainfall seasons have been characterised by media campaigns encouraging the public to change how they use water. Water efficiency education that relies on effective messaging is thus emerging as an influential demand management intervention. An experimental study conducted by ESW in fact showed that customers who received messages relating to behavioural change recorded more water savings (by 7 l/property/day) than those who did not receive any message (Ross, 2015).

4.3.3 Combined water efficiency engagement

More widely, water utilities in countries belonging to the Organisation for Economic Co-operation and Development (OECD) now complement technical measures with educational interventions to achieve water efficiency (Grafton *et al.*, 2011: 2). This is in essence the combined water efficiency engagement approach.

The compound water savings that can be realised from techno-efficiency and edu-efficiency have inspired increased calls for combined water efficiency engagement in the UK water sector (see Waterwise, 2015). Waterwise's 2015 report demonstrates that water efficiency home visits are by far the most popular type of initiative used by UK water utilities to deliver combined water efficiency engagement on the frontline. In doing so, water efficiency home visits centre around installing water saving devices in households and 'educating' residents about water efficiency.

The challenge with combined water efficiency engagement, however, is that water utilities are critiqued by practice theorists as merely tinkering with water efficiency education. Combined water efficiency engagement is portrayed to be simplistic whereas it can be indeed complex because diverse usage practices mean that utilities tend to engage with people flexibly when using this approach. But water utilities' narrow tactics for educating people about using water wisely render the maximisation of water efficiency challenging. For instance, home visit campaigns still tend to focus more on how devices can assist people to reduce usage and the financial savings that bill-paying customers can make as a result, rather than equally focussing on addressing the complex factors that impact water efficiency such as institutions, attitudes, norms, resources, technology, and water systems. In doing so, these home visit campaigns fail to consider water efficiency as an outcome of public engagement, water management practice, and other externalities as much as it is an outcome of the end use of water. There is therefore a need for practitioners to look at how they engage with the public about water efficiency and seek opportunities to reconfigure day-to-day strategies for encouraging people to think differently about water and their usage. This brings to the fore the relevance of sociotechnical water efficiency engagement discussed in the next subsection.

4.3.4 Sociotechnical water efficiency engagement

Discussions about the sociotechnical approach to water efficiency engagement is new in the demand management literature compared to the other approaches. Sociotechnical water efficiency engagement draws on practice theory which locates water usage within a social and material context (see Browne *et al.*, 2013) and advocates robust tackling of water demand by targeting the multiple socio-material factors that influence patterns of water usage such as norms, values, resources, socio-economic conditions, institutions, environment, technology, and water systems (see Watson *et al.*, 2020).

Sociotechnical water efficiency engagement factors in internalities and externalities that shape patterns of water usage, for example, people's normative beliefs (e.g., linking cleanliness to laundering or showering), age, individual values, awareness about water, communal values, the nature and extent of water efficiency support utilities provide to people, people's level of trust in their water utility, lived experiences of water restrictions, home ownership, garden ownership, and the presence or absence of efficient water systems in households and so on (see Ajia, 2021). If executed effectively, this approach could bring about a reflexive multi-stakeholder co-production of water efficiency. However, whilst traces of this approach can be found in the UK water industry (e.g., liaisons seen between some water utilities and local authorities and housing associations to enhance water efficiency in social houses), sociotechnical water efficiency engagement is still an aspiration due to its newness.

The next section introduces ESW's EDC home visit campaign as an illustration of water efficiency engagement in practice, including barriers faced on the frontline and practice improvements made during the last AMP6 period, and the state of the matter as the utility is keeping engagement going during the current AMP7 period despite the challenges faced due to the COVID-19 pandemic.

4.4 WATER EFFICIENCY EDUCATION IN ESW

The impacts of climate change and population growth seen in the UK between the 1990s in the forms of low reservoir levels, drought events, and increased water usage motivated ESW to re-evaluate its action plan to increase household water efficiency and create its water efficiency team in 1997. At this time, the water efficiency team distributed water saving packs to households and self-installation of these devices was advised to encourage customers to take responsibility for measuring and reducing their usage. However, the utility's quick recognition that water savings from self-installation of water saving packs had plateaued led to a decision that water efficiency needed to be promoted on a more robust and systematic scale. It was in this light that the H2eco project was developed in 2007, and then redeveloped and rebranded as the EDC campaign in 2015.

The EDC campaign is a novel annual catchment-based initiative forming a key part of ESW's new water efficiency strategy. This water efficiency strategy was to contrast the self-installation approach to supporting the public to retrofit their water systems which was seen decades ago. The EDC campaign seeks household water efficiency via retrofitting and direct public engagement and advice-giving to change how people use water in the home. More broadly, the campaign draws upon wider perspectives that locate water efficiency within the behaviour change context, and its engagement activities seek to target habits relating to patterns of water usage. Whilst the face-to-face water efficiency home visit is the staple of the EDC campaign, it must be noted that the water utility also delivers plays and workshops in schools and holds gardening events and other awareness events in its catchment area to promote key water efficiency messages. Marketing campaigns are also held in public spaces such as town centres to give water efficiency advice and encourage members of the public to sign up for a home visit. However, these peripheral activities are not the primary focus of this book chapter.

In the summer of every year, qualified plumbers visit households registered for a home visit in a particular town to audit their properties, retrofit their water systems, and have educative conversations with residents about positive behaviours and practices that can yield water savings. Thus, in line with the focus of this chapter, the next subsection reflects on the barriers faced and practice improvements that were implemented by ESW to increase resilience as the water efficiency team and the plumbers worked to deliver home visits during AMP6 (2015–2020).

4.4.1 The home visit campaign during AMP6: barriers faced and practice improvements made

During AMP6 (2015–2020), the EDC home visit campaign was designed and delivered to maximise multiple demand management interventions. In what can be seen as a combined approach to water efficiency engagement, strategies for home visits centred around retrofitting water systems and encouraging people to adopt positive behaviours around water use whilst also promoting the installation of the normal water meter, if necessary, to increase its uptake and maximise water savings.

The water efficiency home visit campaign was delivered in Leigh-On-Sea in 2015, in Lowestoft in 2016, in Witham in 2017, in Barking in 2018, and again in Leigh-On-Sea in 2019 (see Figure 4.1).

A typical home visit during AMP6 spanned 45–60 minutes. On arrival at the customers' homes, the plumbers would introduce themselves and then discuss the aim of the home visit with the residents, drawing upon an engagement script which they had been trained to use. Following a safety assessment, the plumbers would read the water meter if the property has one, and then conduct a water efficiency audit. This audit would entail assessing the property to determine which water saving devices can be installed, after which the plumbers would then retrofit water systems wherever necessary whilst also educating available residents about water efficiency.

In retrospect, significant effort went into the fusion of face-to-face water efficiency education with technical water efficiency engagement during AMP6 in ESW. However, invaluable lessons were learnt from the barriers to water efficiency engagement faced during the annual home visit campaigns. And the water efficiency team saw these barriers as opportunities to improve and maximise engagement techniques on the frontline. These practice improvements concerned: (1) linking household space, instruments of change, and messaging during engagement; (2) raising the stake of the behavioural change education aspect of home visits; (3) enhancing the plumbers as a channel of communication; and (4) improving customer insight development and maximising its use. These practice improvements will now be discussed

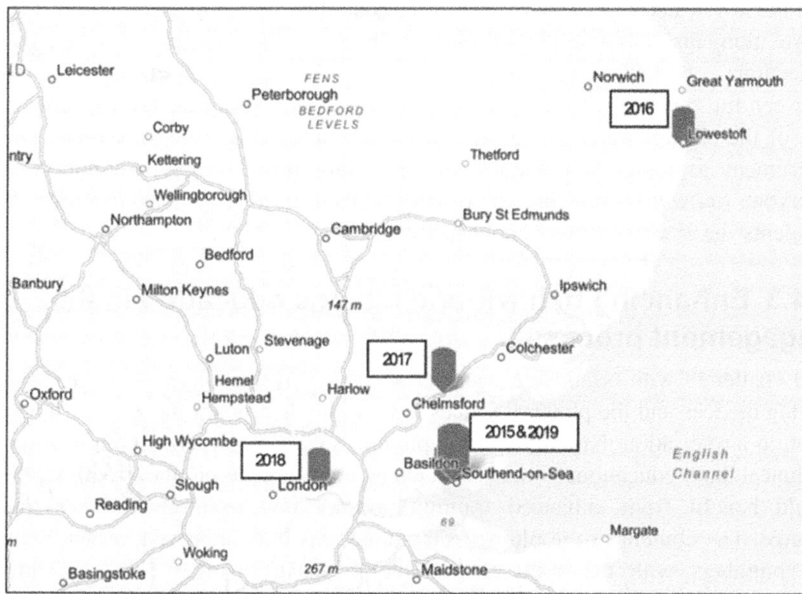

Figure 4.1 Areas where the EDC water-efficiency home visits were delivered during AMP6. *Sources*: Generated by authors.

collectively as they are interconnected in the way they impact the quality of water efficiency engagement.

4.4.2 Maximising the links between household space, water saving devices, and water efficiency messaging

It emerged that the plumbers had a peculiar pattern of movement in and around spaces whilst auditing households, following a set order: retrofitting in the kitchen, cloakroom, toilet(s) and the bathroom(s), and other spaces within the premises such as the garden (Ajia, 2021). Typically, whilst the plumbers were checking for leaks in the kitchen, toilets, and bathrooms, they would check taps in the kitchen and garden; and they would audit cistern(s) and taps in toilets and bathrooms. Upon completion of home audits, the plumbers would give residents non-plumbing water saving devices such as the trigger hose gun, water crystals, bath buoy, plate scraper, and shower timer, and a water butt if required and necessary. Then, the plumbers would also give the residents an information pack containing literature about behavioural change and a product guarantee card to conclude their home visit.

The challenge however was that the plumbers did not always actively *follow the space* in making the connections between the water saving devices being retrofitted and the key water efficiency messages that could be powerfully communicated in those moments. Observation of the plumbers on the frontline and their own reflections on their engagement techniques revealed some disorganisation between the physical space, water saving device, and message. For example, the use of the shower timer to 'excite' customers rather than being discussed as an instrument to nudge a reduction in showering time, or the assumption that everyone knows the function of a dual-flush rather than probing to understand residents' level of awareness about its use.

4.4.3 Enhancing behavioural change education in the engagement process

The challenge with organising water efficiency conversations with related water saving devices and the physical spaces in households brought the water efficiency team to a realisation that although the plumbers were delivering a combination of technical and educational water efficiency engagement on the frontline, they could benefit from enhanced training around how to motivate behavioural change. The content of the plumbers' training was thus improved to better equip the plumbers with the skills to be able to make linkages between various elements that influence household water usage such as norms, technology, water systems, knowledge, and values. The plumbers were also trained to hold water conversations with residents, whilst concurrently auditing, repairing, and retrofitting water systems, and installing water saving devices, as well as recording installation data (such as measurements, flow rates, leak details, and before and after photographs of retrofitted water systems) in their personal digital assistant (PDA) devices.

4.4.4 Maximising plumbers as a channel of communication

Observation of the plumbers on the frontline and their own reflections on their engagement techniques revealed that they were more confident about the technical aspects of the home visit than about the education aspect of the water efficiency engagement. This is perhaps not surprising for a set of people who have a technical background, with little explicit training in relation to in relation to the educational aspects of their role. The role of the plumbers in the achievement of household water efficiency and the need to maximise this human asset therefore came to the forefront during AMP6 more than ever before.

Considering that existing studies in the environmental literature position personnel on the frontline as a channel of communication (see Mony, 2007; Mony & Heimlich, 2008), the water efficiency team recognised that the plumbers were carriers and influencers of water efficiency messages and engaged with academia to provide them with social science-led training.

The plumbers' training was improved to equip the plumbers with the ability to use communication techniques to keep household residents within proximity during home visits engaged and encouraged to actively participate in their home visit experience. For example, the format of the plumbers' training was expanded to include role playing so that they could enact and prepare for encounters on the frontline.

Further, a practical aspect of the water efficiency home visit campaign that was inculcated and was increasingly emphasised during the plumbers' training in AMP6 was behavioural change education that is not just about information provision and product demonstration but also dialogue. Of course, information sharing ensures that household residents are aware of the reasons for water saving device installations and retrofits. And such information sharing helps mitigate the risk of a resurging high water usage that may occur when residents lack knowledge about their water efficiency interventions. However, the role of information sharing during home visits in ESW was further enhanced to include dialogue and feedback so that on the part of residents, their appreciation for water increased and on the part of the plumbers, customer insight was gained for the purpose of practice improvement. This reimagining of information sharing during home visits fostered lesson-learning and has increased the co-creation of practice improvement by the plumbers and the water managers which will be discussed in the next subsection.

4.4.5 Customer insight development and use in practice improvement

The water efficiency home visit campaigns during AMP6 exposed how the plumbers were a trove of customer insight for ESW, and how working with the plumbers could help the water utility further improve its practices.

The home visit data recorded in the PDA devices usually centred around the plumbers' arrival and completion times, the size of the household, the number of toilets and bathrooms, the reason(s) why water saving devices were not installed, and photographs of water systems before and after retrofitting and so on. However, whilst these aforementioned data are useful, they provide little understanding of customers' water values, behaviours, and usage.

Although light-touch customer insight came from the customer satisfaction survey administered to households seven weeks after their home visit, it did little to advance the utility's understanding of its customers. The customer satisfaction survey aimed to understand customers' interaction with their newly installed water saving devices, test the effectiveness and sufficiency of the information in their information pack, and understand the unmeasured impact of home visits.

The study conducted by Ajia (2021) revealed that the plumbers were privy to invaluable qualitative customer insight information which was not being captured in the PDA devices, and hence remained unknown to water managers. For example, whilst the plumbers would know who had a garden and may be a suitable invitee for other gardening events to enhance household water efficiency, such information was not being passed on to the water efficiency team. This is because the plumbers were far removed from designing and project-planning the water efficiency home visit campaign. This barrier was however addressed in real time during the delivery of the 2017 water efficiency home visits in Witham. That year, the water efficiency team created a plumbers' forum to bridge the communication gap between the water managers and the plumbers and this has led to increased feedback and collaborative work between the parties, enhancing the extent of customer insight for practice improvement. This is because the plumbers' forum has offered a regular informal setting for the plumbers and the water managers to share knowledge, reflect on practices and lessons learned on the frontline, and (re)design engagement processes throughout the life cycle of the EDC home visit campaigns.

Having discussed some of the practice improvements that ESW made to water efficiency engagement on the frontline during AMP6, the next section discusses new challenges that have emerged at the beginning of AMP7 (2020–2025) considering the COVID-19 pandemic and pragmatic adaptations being made to the utility's water efficiency engagement approach to maintaining resilience to water scarcity.

4.5 THE HOME VISIT CAMPAIGN DURING AMP7: NEW CHALLENGES AND ADAPTATIONS

ESWs delivery of water efficiency education in AMP7 (2020–2025) has commenced atypically as the world is going through unprecedented times with the COVID-19 outbreak. In March 2020, the water utility paused the planning of the 2020 EDC home visit campaign due to the spread of COVID-19 and the need to put health and safety first. The early pause of the 2020 EDC home visit campaign meant that no arrangements for face-to-face water efficiency engagement were initiated with customers at all, and other demand reduction interventions such as fixing leaky loos in homes and face-to-face water efficiency education in schools were suspended temporarily.

4.5.1 New challenges due to the Covid-19 pandemic

WaterBriefing (2020) reported that in 2020, the UK saw the driest May since records began. The hot summer season also coincided with the first COVID-19 lockdown and widespread public health messages promoting frequent handwashing and general cleanliness to help prevent the spread of COVID-19. Considering that more people were spending more time at home and increasing the frequency of their water usage activities, it is no surprise that the water industry saw an increase of 20–40% in household water use (Water UK, 2020).

A preliminary survey conducted by Essex & Suffolk in July 2020 revealed that the number of customers working from home increased from 7.45% before the first

COVID-19 lockdown to 28.31% after the lockdown. It is thus expected that water usage practices such as cooking, flushing, dishwashing, showering, handwashing, gardening, use of hot tubs, paddling pools and swimming pools would have compounded the demand–supply imbalance faced due to the dry summer and impacted on customers' water bills. And low rainfall meant that whilst outdoor water usage increased, water harvesting using water-butts reduced.

Therefore, whilst a conclusion about the level of increase in per capita consumption particularly in Essex & Suffolk could not be drawn at the time of writing (March 2021), the assumption can be made that household water usage has increased due to transfer of usage from external non-residential spaces to residential spaces.

It must be noted that ESW already had established preparedness measures for meeting increased demand during exceptionally hot and dry summers, and the water utility's leaky loo intervention to repair leaking toilets resumed towards the end of 2020. Nevertheless, it is still imperative that the water utility continues to seek creative and new ways to mobilise the public to reduce their water usage to relieve pressure on water resources and considering that more customers are requesting support to understand why their water bill is rising or to obtain bill payment holidays. Thus, in addition to maximising other alternative modes of customer engagement, such as social media, the water utility launched its virtual water efficiency home visit campaign in September 2020.

4.5.2 The virtual water efficiency home visit campaign

A typical virtual home visit is delivered via video conference (Figure 4.2) on a safe and secure platform. The development of the virtual water efficiency home visit campaign was a collaborative effort between contractors and some internal business functions including the water efficiency team, procurement, marketing and communications, and the systems team. Like the face-to-face home visit campaign, the aim of the virtual home visit campaign was to increase household water savings. The campaign was piloted (from September 2020 to October 2020) in 66 households in 2 rural areas of Suffolk and participants were recruited via email invitation and follow-up phone calls, representing a 3.78% uptake rate.

To commence the home visit process with the plumber at a scheduled time, the customer (usually the bill-payer who agreed to the visit) clicks on a dedicated link which would have been emailed to them previously. Like the face-to-face home visit, the virtual home visit relies on the participation of the customer to conduct an audit of the property, provide tailored water efficiency advice, and identify water saving devices suitable for the water systems in the property such as taps, toilets and showers. Then, the identified water saving devices are posted to the customer to self-install. Whilst the virtual home visit is not an entire departure from the physical home visit experience, distinctions can be inherently found in

Figure 4.2 An illustration of the virtual water efficiency home visit. *Source*: Aqualogic.

its format which causes increased customer involvement during and after engagement has taken place. This is because the plumber is very reliant on the customer to show them around the house and taking water meter readings and self-installation of water saving devices are ultimately up to them – the customer. Although customers are provided with an option for a follow-up video call to be assisted with their self-installation, none of the participants during the pilot of the virtual home visit campaign requested to have one. And follow-up contact with the participants revealed that they were all able to self-install easily.

4.5.3 How success of the virtual water efficiency home visit pilot campaign was measured

ESW does not currently have smart water meters. The calculation of measured water savings is therefore based on manual readings of water meters taken before and after home visits. Due to the COVID-19 pandemic and lockdown restrictions, the water utility did not visit households to take water meter readings during the virtual home visit pilot campaign but relied on householders to do so independently and provide their readings to the utility. There was a low submission rate (for meter readings) as only 10% of participating households provided readings before self-installing their water saving devices. As a result, it was not possible to accurately determine the measured water savings yielded from the virtual home visits.

The level of success of this pilot campaign was determined based on the number of water saving devices posted to households. The newness of the virtual home visit campaign means that opportunities to maximise and improve this form of water efficiency engagement are emerging and the ways to measure effectiveness are still developing. Going forward, the water utility will improve the reconciliation between the aim of the virtual home visit campaign and how performance is measured, and it will strengthen its liaison with householders to determine measured water savings.

Nevertheless, the average number of water saving devices fitted per property during the virtual home visit campaign (4.4) was lower than the average fitted per property during face-to-face home visits (6.4) (Figure 4.3).

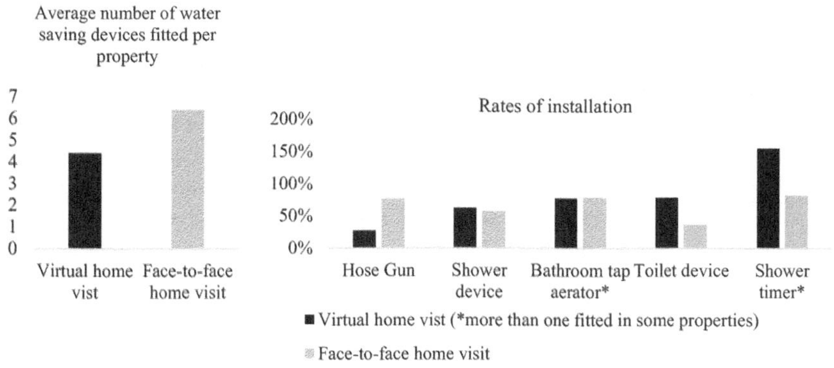

Figure 4.3 Comparison of rates of fitting of water-saving devices between home visit types. *Source*: Generated by authors.

Whilst Figure 4.3 shows similar rates of installation for shower and bathroom devices during the virtual and face-to-face home visit campaigns, it must be noted that the fixing of leaky taps and some water saving devices (e.g., the ecoBETA device for retrofitting single-flush toilets and shower heads) were not provided to properties during the virtual campaign. Also, the rates of installation for the virtual campaign were estimated based on the number of devices posted to properties, not necessarily confirmed as self-installed in properties. This contrasts the rates of installation for the face-to-face campaign which are accurate since the water utility installed the water saving devices.

In view of the above and based on historic data which suggests that face-to-face water efficiency home visits during AMP6 delivered an average measured water saving of 21 litres/property/day, the conclusion can thus be drawn that the unmeasured water savings realised from the virtual water efficiency home visit pilot campaign were lower than the measured water savings from the face-to-face home visit campaign.

4.6 LESSONS LEARNED

Venturing into virtual water efficiency home visits has caused ESW to think more creatively about how water efficiency support can be better provided to the public. The lessons distilled from the pilot campaign include the following eight points:

- Although pausing the face-to-face home visit campaign meant that ESW lost some of the social benefits conventionally gained from delivering peripheral community outreach and marketing events, increasing social media engagement, direct emails, letters, phone calls and text messages to customers has helped the water utility to maintain visibility. Also, the occurrence of COVID-19 has increased the public's awareness of water. Work is therefore underway to find ways to sustain customers' engagement with water efficiency and the water utility on a long-term basis.
- Recruitment of households for the virtual home visit pilot campaign via email was relatively ineffective. The water utility is therefore exploring other recruitment channels to improve the uptake rate for the future.
- Virtual water efficiency home visits contrast the water efficiency engagement that is normally seen in the annual face-to-face home visit campaign which is season-based and town-based. For the future, customers in ESWs catchment area could be engaged regardless of time and place. Given that the water utility has an ambitious goal to engage with every single customer (at least two million participating by 2025), virtual home visit campaigns could increase customer reach through water efficiency engagement.
- On the part of the water utility, virtual water efficiency home visits have led to reductions in travel time to conduct water efficiency home audits, financial and environmental costs, and carbon footprint. These have helped increase ESWs adaptation to uncertainty and shock.
- On the part of the customers, virtual home visits have afforded them greater flexibility as well as increased their involvement in and responsibility for water efficiency. ESW however recognises that visiting households to support retrofitting or to take meter readings ultimately defeats the purpose of the virtual home visit campaign. The effective execution of virtual home visits in the future should therefore include increased support to customers to self-install a wider range of products and to take meter readings independently.
- There was a high uptake of water saving devices to support customers to shower for shorter durations. Work is therefore underway to improve the design and delivery of water efficiency messages around this water usage activity to maximise water savings.
- There was a quick realisation that virtual home visits should not preclude face-to-face home visits and vice versa. Going forward, the virtual and face-to-face home visits will serve as a twin-track strategy for the flexible delivery of water efficiency engagement in ESW.

- Feedback from the virtual home visit pilot campaign suggests that the campaign was satisfactory for supporting customers to determine product suitability and know more about the water saving devices that will be posted to them. However, more could be done to increase the water efficiency education aspect of the campaign. Going forward, ensuring that conversations and dialogue about water efficiency and residents' water usage behaviours occur as a complementary rather than supplementary aspect of virtual home visits will be a priority. In addition, the development and use of insights from water efficiency education during virtual home visits will be prioritised.

ESWs next and immediate plan is to deliver 1350 more virtual home visits to households in Suffolk and more broadly in the water utility's catchment area before the end of March 2021. The water utility's long-term ambition beyond 2025 is to meet the National Infrastructure Commission's target to reduce per capita consumption to 118 l/p/d by 2040 (NIC, 2018). It is therefore crucial that the water utility implements follow-on actions from lessons learned and continues to improve its mobilisation of the public to increase household water efficiency.

The next section brings this book chapter to a close by consolidating how innovation in ways of working on the part of the water utility can contribute to the advancement of water efficiency engagement as a practice.

4.7 CONCLUSION

There is a water scarcity crisis in the UK and water utilities are under pressure to find new and innovative ways to reduce household water usage. During the previous AMP6 period (2015−2020), ESW sought to achieve both techno-efficiency and edu-efficiency. It came to the fore that to advance water efficiency engagement, there is the potential to begin moving towards sociotechnical change. It is beneficial to upskill plumbers in the aspect of water efficiency education; engage with academia during project-planning; and maximise the linkages between staple and peripheral water efficiency activities. Further, the occurrence of the COVID-19 pandemic as the water industry begins delivering the current AMP7 (2020−2025) has shown that uncertainties and shocks, however disruptive, can motivate innovative practice improvement. Emerging ways to further water efficiency engagement as a practice thus include the delivery of virtual water efficiency home visits side by side with face-to-face home visits, and the improvement of the quality of complementary water efficiency education delivered with retrofitting on the frontline. Flexible virtual home visits that are not season-bound or town-specific address the time and place boundaries that accompany face-to-face home visits. This can open opportunities to engage new segments of the public such as working families and rural area dwellers. But it must be noted that virtual water efficiency engagement can exclude other segments of customers such as other household occupants besides the bill-payer and bill-paying customers who do not have access to or choose not to use online technologies for engagement.

REFERENCES

Adaptation Sub-Committee (2016). UK Climate Change Risk Assessment 2017 Synthesis Report: Priorities for the Next Five Years. Sub-Committee of the Committee on Climate Change, London, United Kingdom.

Ajia F. O. (2018). Examining adaptation using the message actor channel (MAC) model of communicative water practices. *Water Science and Technology: Water Supply*, **18**(4), 1318–1328.

Ajia F. O. (2021). Water Efficiency Engagement in the UK: Barriers and Opportunities. PhD thesis, Sheffield Water Group, The University of Sheffield, UK.

Browne A., Med W. and Anderson B. (2013). Developing novel approaches to tracking domestic water demand under uncertainty – A reflection on the top 'Up scaling' of social science approaches in the United Kingdom. *Water Resource Management*, **27**(4), 1013–1035.

Browne A. L., Jack T. and Hitchings R. 2019. 'Already existing' sustainability experiments: lessons on water demand, cleanliness practices and climate adaptation from the UK camping music festival. *Geoforum*, **103**, 16–25.

Department for Environment Food and Rural Affairs (2018). Water Conservation Report. Available at https://assets.publishing.service.gov.uk/government/uploads/system/ uploads/attachment_data/file/766894/water-conservation-report-2018.pdf (accessed 29 September 2020).

Dworak T., Berglund M., Laaser C., Strosser P., Roussard J., Grandmougin B., Kossida M., Kyriazopoulou I., Kolberg S., Montesinos Barrios P. and Berbel J. (2007). EU Water Saving Potential (Part 1-Report), Report ENV.D.2/RTU/2007/001r, Institute for International and European Environmental Policy, Berlin, Germany.

Grafton R. Q., Ward M. B., To H. and Kompas T. (2011). Determinants of residential water consumption: evidence and analysis from a 10-country household survey. *Water Resources Research*, **47**(8).

Hamling I., Bloomfield W., Dearing K. Y. and Watson T. (2018). Optimising demand reduction in water utilities. *EPiC Series in Engineering*, **3**, 874–883.

Keeting T. and Styles M. (2004). Performance Assessment of Low Flush Volume Toilets: Final Report for Southern Water and the Environment Agency.

Knamiller C., Medd W., Sefton C. and Sharp E. (2006). Heybridge 2006: Water Efficiency Technology in Everyday Life and People's Perception of Personal Use, Report for Essex & Suffolk Water.

Koop S. H. A., Van Dorssen A. J. and Brouwer S. (2019). Enhancing domestic water conservation behaviour: A review of empirical studies on influencing tactics. *Journal of Environmental Management*, **247**, 867–876.

Lawson R., Marshallsay D., Difiore D., Rogerson S., Meeus S. and Sanders J. (2018). The Long-Term Potential for Deep Reductions in Household Water Demand. Available at https://Www.Ofwat.Gov.Uk/Wp-Content/Uploads/2018/05/The-Long-Term-Potential-

For-Deep-Reductions-In-Household-Water-Demand-Report-By-Artesia-Consulting. Pdf (accessed 20 June 2018).

Mony R. S. P. (2007). An Exploratory Study of Docents as a Channel for Institutional Messages at Free-Choice Conservation Education Settings. PhD thesis, The Ohio State University, Ohio, United States.

Mony P. R. and Heimlich J. E. (2008). Talking to visitors about conservation: exploring message communication through docent–visitor interactions at zoos. *Visitor Studies*, **11**(2), 151–162.

National Infrastructure Commission (2018). Preparing for a Drier Future: England's Water Infrastructure Needs. National Infrastructure Commission, London, United Kingdom.

Office of National Statistics (2018). Available at https://www.ons.gov.uk/people populationandcommunity/populationandmigration/populationestimates/articles/over viewoftheukpopulation/november2018 (accessed 25 January 2019).

Office of Water Services (2011a). Push, Pull, Nudge. How can We Help Customers Save Water, Energy and Money? Ofwat, London, United Kingdom Available at https:// webarchive.nationalarchives.gov.uk/20150604064034/http://www.ofwat.gov.uk/publi cations/focusreports/prs_web1103pushpullnudge (accessed 04 September 2019).

Office of Water Services (2011b). Involving Customers in Price-Setting – Ofwat's Customer Engagement Policy, Ofwat, London, United Kingdom. Available at http://webarchive. nationalarchives.gov.uk/20150624091829/https://www.ofwat.gov.uk/future/mono polies/fpl/customer/pap_pos20110811custengage.pdf (accessed 06 March 2017).

Orr P., Papadopoulou L. and Twigger-Ross C. (2018). Water Efficiency and Behaviour Change Rapid Evidence Review, Joint Water Evidence Programme, Final report WT1562, project 8, DEFRA, London, United Kingdom.

Owen L., Bramley H. and Tocock J. (2009). Public Understanding of Sustainable Water Use in the Home: A Report to The Department for Environment, Food and Rural Affairs, Synovate, DEFRA, London, United Kingdom.

Parliamentary office of Science and Technology (2012). Water Resource Resilience. POSTNOTE Number 419. POST, London, United Kingdom.

Ross J. H2eco behavioural research (Phase 10), Mouchel, London, United Kingdom. Available at H2eco-Research-Phase-10-Final-Report.pdf (waterwise.org.uk) (accessed 18 September 2019).

Smith S. and Shouler M. (2001). Sustainable New Homes, Heybridge, Essex. In: 2006 Water Demand Management, D. Butler and F. Memon (eds), International Water Association Publishing, London.

UK Environment Agency (2021). Water Stressed Areas – Final Classification 2021, Environment Agency, Bristol, United Kingdom. Available at https://view.officeapps. live.com/op/view.aspx?src=https%3A%2F%2Fassets.publishing.service.gov.uk%2Fg overnment%2Fuploads%2Fsystem%2Fuploads%2Fattachment_data%2Ffile%2F9982 37%2FWater_stressed_areas___final_classification_2021.odt&wdOrigin=BROWSEL INK (accessed 21 July 2016).

UK Environment Agency and Natural Resources Wales (2013). Water Stressed Areas – Final Classification, Environment Agency and Natural Resources Wales, London & Bristol, United Kingdom. Available at https://www.gov.uk/government/uploads/system/ uploads/attachment_data/file/244333/water-stressed-classification-2013.pdf (accessed 21 July 2016).

WaterBriefing (2020). Demand for Water Surges Due to Hot Weather and Coronavirus Lockdown. Available at https://www.waterbriefing.org/home/water-issues/item/172 78-demand-for-water-surges-due-to-hot-weather-and-coronavirus-lockdown (accessed 04 November 2020).

Water UK (2020). Be 'water aware' – Tips for People at Home. Available at https://www.water.org.uk/news-item/be-water-aware-tips-for-people-at-home/ (accessed 01 December 2020).

Waterwise (2013). Water Efficiency and Drought Communications Report. Available at http://www.waterwise.org.uk/wp-content/uploads/2018/01/201McKenzie-Mohr3_Waterwise_Drought_Report.pdf (accessed 11 August 2018).

Waterwise (2015). Water Efficiency Today: A 2015 UK Review. Available at https://www.waterwise.org.uk/wp-content/uploads/2018/02/Water-Efficiency-Today-UK-Review- 2015.pdf (accessed 06 November 2018).

Watson M., Browne A., Evans D., Foden M., Hoolohan C. and Sharp L. (2020). Challenges and opportunities for Re-framing resource Use policy with practice theories: The change points approach. *Global Environmental Change*, **62**, 102072.

Chapter 5

Water resources east: An integrated water resource management exemplar

Nancy Smith[1], Robin Price[2] and Steve Moncaster[3]

[1]*Communications and Engagement, UK, Water Resources East, Norwich, United Kingdom*
[2]*Managing Director, UK, Water Resources East, Norwich, United Kingdom*
[3]*Technical Director, UK, Water Resources East, Norwich, United Kingdom*

ABSTRACT

Water Resources East (WRE) is a 180 strong and growing membership organisation established in 2014 to learn from international best practice on how to develop a more collaborative approach to water resource management planning to the 2050s and beyond. This is happening now in a unique region of England under significant pressure due to population growth, economic ambition, the need for enhanced environmental protection, and significant climate change impacts.

The lesson of this chapter is the power of multi-sector water resource planning through collaborative and adaptable mechanisms led by integrated water resource management (IWRM). Through using active project case studies to gain insight into how we work with our members: Future Fenland Adaptation; Regional Natural Capital Planning through Systematic Conservation Planning (Water Resources East is teaming up with Biodiversify and WWF-UK, with financial support from the Coca-Cola Foundation, to develop a natural capital plan for Eastern England through Systematic Conservation Planning); and exploration of multi-sector finance of nature-based solutions through the creation of Water Funds, we hope to provide a strong evidence base for our sustainable and resilient methodologies and approaches that can be used, or be an influence on, other water management systems globally.

doi: 10.2166/9781789061628_0081

Lastly, the WRE team and longest standing contributors reflect on lessons and recommendations from the past seven years of work.

Keywords: water scarcity, stakeholder engagement, integrated water resources management, catchment-based approaches.

5.1 INTRODUCTION

Eastern England is unique and characterised by its low rainfall, internationally important habitats and diversity of water use. As such, the focus of Water Resources East (WRE) is on multi-sector water resource planning with a vision for the region to have sufficient water resources to support a flourishing economy, a thriving environment, and the needs of its population.

At the time of writing WRE has over 180 members including public water supply; energy; Internal Drainage Boards; landowners and farming representatives; regional governments; environmental & conservation groups; community and advocacy groups; university and education institutions; and private businesses.

It is crucial that all of these organisations are involved in a collaborative and integrated planning approach due to the diversity of water usage in the region and the risk factors attributing to uncertainty over future availability. In the UK Environment Agency's National Framework for Water Resources (2020) they state that 85% of water taken from the environment in Eastern England is used for public water supply, compared with around 98% in London and the South East of England. Of the remaining 15%, well over half is used to irrigate crops as our region contains some of the most productive agricultural land in the country. There is also significant, and growing water use by the energy and wider industrial sectors.

It is estimated that if no action is taken by 2050 the regional gap between supply and demand could be 570 Ml/day (megalitres per day) for public water supply and 444 Ml/day for agriculture, power, and industry (*ibid*).

Set against this backdrop is the crucial requirement for environmental restoration and enhancement in the region, with a clear recognition that ensuring a thriving water environment will be key to securing a thriving economy.

As an independent legal entity (a Company Limited by Guarantee) Water Resources East (WRE) Ltd. now operates as an inclusive, collaborative, exploratory and forward-thinking membership organisation, charged with the development of a long-term Regional Plan to ensure that the region can meet both its environmental and economic ambitions in the context of a changing climate.

5.2 OUR CONTEXT (PRICE & WRE, 2020)

In its 25-year Environment Plan (2018), the UK Government pledged that we would be the first generation to leave the environment in a better condition than we found it.

To help meet the pledge to improve resilience to drought and minimise interruption to water supplies, the Environment Agency, one of five government water sector regulators, has led the development of a National Framework for Water Resources in England which was published in March 2020.

The National Framework *(ibid.)* evidences the strategic long-term water needs of England, both nationally and within the boundaries of the regional water resources groups. It does this for all sectors that depend on a secure supply of water while also ensuring the environment is improved. The National Framework, while led by the Environment Agency, has been developed in collaboration with the other regulators – Ofwat and the Drinking Water Inspectorate (DWI), and the Department for the Environment, Food and Rural Affairs (Defra), as well as a wide range of stakeholders represented by a senior steering group made up of around 40 water industry representatives, other water users, environmental non-governmental organisations (NGOs), government and regulators from England and Wales.

The National Framework *(ibid.)* is part of the water resources planning cycle. Five regional groups now exist across England, and the National Framework sets the challenge for these regional groups to work collaboratively to develop ambitious regional water resources plans that provide resilient and efficient water supplies into the future and that have environmental enhancement at their core. Regional groups, of which WRE is one, are critical to the development of integrated plans that include the right strategic solutions for the challenges facing the nation, and each regional group has been tasked with pulling together a single multi-sector integrated water resource management plan.

WRE is currently developing a regional integrated Water Resources Management Plan (the Regional Plan) covering catchment areas across the East of England and part of the East Midlands (herein after referred to as 'Eastern England'). As a group we aim to co-create and build a long-term, multi-sector adaptive plan that reflects the needs and characteristics of our diverse region.

For the WRE region, this plan will:

- Seek to increase the level of resilience for water resources for all sectors.
- Identify opportunities to deliver wider benefits in terms of flood risk and water quality.
- Identify ways to ensure that water (either too much or not enough) is not a barrier to economic development in the region.
- Seek to enhance the environment, in line with the 25 Year Environment Plan.
- Explore innovative funding and delivery models for water management solutions.
- Promote schemes which represent the best value for the region, seeking through collaboration to deliver more efficient solutions.
- Co-deliver the water-related elements of other key regional strategies and plans.

- Focus on delivery of water-related climate change mitigation and adaptation strategies including net zero carbon ambition.
- Provide academically rigorous evidence to policy makers.

A first draft of the regional plan will be published in early 2022. To get to this point, 2021 will consist of planning conferences to co-create more focused portfolio options ready for consultation and detailed national alignment work. The second, more advanced draft of the Regional Plan will be published in August 2022 at the same time as the first draft of Water Resource Management Plans that are required from all public water supply companies. The final draft will be published in September 2023. The Regional Plan will cover a time period from now until the 2050s.

5.3 OUR REGION

Eastern England is a unique and diverse part of the UK. The Norfolk Broads and wetlands of the East Coast are internationally recognised and have more than a quarter of Britain's rarest wildlife with 125 miles of waterways. Three of the UKs five fastest growing cities, Cambridge, Norwich and Peterborough are home to 10.5 million people and growing, all with additional housing needs. The region's far reaching, fertile agricultural lands provide 40% of England's vegetables worth £2.8 billion a year, yet 30% of the land mass is below sea level.

Eastern England is the driest region in the UK and is susceptible to prolonged, unpredictable, droughts, such as that in 2018 (The National Farmers' Union, 2018). It is also susceptible to flooding events with flood management organisations known as Internal Drainage Boards having to pump millions of litres of water out to sea from the region's long coastline each year. The irony being that this is water which could be harnessed to use in times of need.

With the increasing risk of drought and the rise in demand for food, energy and services that is likely in the future, there is a very real risk that a lack of collaborative water management could limit growth and development in our region. The WRE region is predicted to face a significant gap between supply and demand if the region carries on managing water resources the same way as it does now (Figure 5.1).

Given the unique circumstances in Eastern England, WRE had a rare opportunity to help lead, shape and inform thinking in the UK and further afield. This is particularly true when considering the present and future pressures on water resources, and how best to manage demands from intensive agricultural production and food processing sectors and rising a population, together with the potential additional water needs associated with alternative energy sources – such as hydrogen.

In parts of the region, there is evidence that current water abstraction regimes are causing damage to the environment, and work is needed to restore more natural

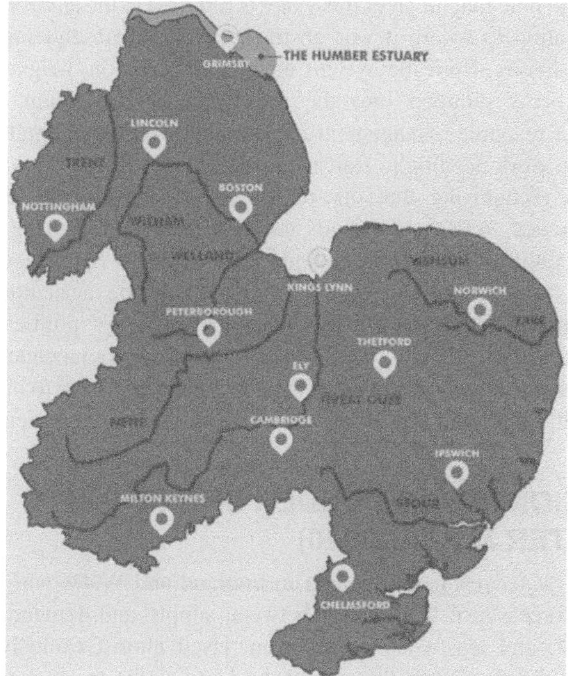

Figure 5.1 Water resources east. *Source*: Price & WRE (2020).

flows in rivers. There is also an assumption that the current pattern of abstraction does not provide an acceptable starting point for planning to meet future needs. Therefore, changes to the current abstraction pattern are needed.

On an average day, in a dry year, 2311 million litres (megalitres) per day are abstracted from the environment and used in Eastern England. Most of this water (85%) is used for public water supply (PWS). Most of the rest is used for spray irrigation (8%), power generation (3%) and in the manufacturing, food and drink sectors (2%).

Abstraction for spray irrigation occurs across the WRE region but is concentrated (71% in terms of licensed volume) to the East of the region. Spray irrigation is strongly seasonal and in a dry year it peaks in July at levels around 600 Ml/d. This is equivalent to approximately 30% of the average daily demand for public water supply.

Abstraction for power generation occurs in the WRE region from the freshwater non-tidal sections of the River Trent and the River Ouse towards the North of the region, from several coastal and estuarine locations, and from a number of the fenland rivers triangulating Peterborough, Ely, and Cambridge. The water is used for cooling and steam generation at coal and gas power stations.

It is correct to note that much of this water is returned to the environment, often at a different location to where it was abstracted, and is subsequently used again. Nevertheless, 'losses' from the system include evaporation, irrigation and water which flows or is pumped into the sea. Therefore, building resilience for improved water resource management for all water users is integral over the next 50 years – with work needing to start now.

As well as recognising that our region is very diverse in terms of water management issues, WRE understands the importance of ensuring that planning is done along political–economic boundaries as well as hydrological ones. The very fact that water does not stop at county borders, agricultural fields, or regulated distribution areas is the very reason why political, economic, environmental, and social aspects need to be considered co-dependently.

Looking to hydrology first, how is WRE's technical approach to integrated water management different, and how does it work in practice as a resilience strategy?

5.4 DECISION MAKING UNDER UNCERTAINTY (MONCASTER & WRE, 2020)

Until recently, water resource planning in England and Wales was dominated by deterministic forecasts of the balance between supply and demand of megalitres per day (Ml/d) and least-cost optimisation. Used almost exclusively by water companies and the regulators, this approach identifies the most cost-effective way to maintain levels of service in a single planning scenario that combines environmental need with best estimates of the future impact of drought, climate change, and population growth. Within this planning framework, risk and uncertainty are accounted for using a planning allowance known as 'Target Headroom'.

Target headroom is an allowance to take into account any uncertainty in the supply demand balance. The basis of the methodology is to apportion the target headroom to two main areas: supply-side and demand-side. An inherent assumption within the methodology is that each component is independent of one another, and where this is not the case risk modelling is used to allow for overlapping, correlated and dependent relationships to be included in the headroom calculation.

While this approach performs well for single sector planning where the supply–demand investment drivers are well understood, and for regions where the predominant use of water is for public water supply, it is less suitable for multi-sector planning or for planning where there is significant uncertainty about investment drivers and the related risks over the long term.

And, as we've seen, Eastern England is different in terms of the amount of water which is used for other purposes, particularly irrigation and power, and in the level of uncertainty into the 2050s.

Therefore, WRE uses integrated water resource management (IWRM) planning that takes account of the uncertainties and risks from many factors including climate change and growth and many sectors. As part of this, WRE has developed a methodology in collaboration with Manchester University that uses a combination of decision making under uncertainty (DMUU) methods, including, multi-objective evolutionary optimisation (MOEO) and robust decision making (RDM).

The MOEO-RDM approach (MO-RDM) allows the vulnerability of water resource systems to be quantified in terms of the impact of growth, climate change, and drought on abstractors from different sectors and the environment. The analysis is simulator based, with uncertainty accounted for by using a wide range of plausible future scenarios, and vulnerability defined in terms of metrics and thresholds which are specified by each sector. Subsequently, MO-RDM identifies 'pareto-optimal' portfolios of schemes that are capable of meeting minimum performance thresholds over a wide range of plausible future scenarios.

In these, performance in respect of one metric cannot be improved unless at the expense of another, therefore trade-offs between the portfolios must be used to select the one which best meets the overall needs of the planners. In this way, WRE can produce strategies and plans which simultaneously meet the needs of the public water supply, environment, energy, and agri-food sectors.

In the last step of the MO-RDM process, the selected portfolio is rigorously stress tested and the vulnerability analysis updated. It is an adaptive process, however, and where additional improvements are needed alternative portfolios can be selected and tested and, if necessary, the process can be re-run based on new information that becomes available.

5.5 STRATEGIC CONTEXT AND IMPLICATIONS

The strategic context for WRE's first regional water resource management plan includes the following:

- Expected but uncertain impacts from climate change on drought and flood risk. Arising from a combination of warmer wetter winters and hotter drier summers, these are likely to be significant and to include the threat of coastal inundation in large parts of the WRE region as a consequence of sea-level rise.

- The commitment to achieve net zero carbon by 2050, and the need to increase levels of economic growth following our departure from the European Union. This includes delivering on the clean growth agenda which is set out in the Governments Industrial Strategy (2017) and 'levelling-up' of disadvantaged communities and areas.

Within the WRE region, more effective integrated water management is absolutely pivotal to meeting these challenges. Securing economic growth and the related benefits for our communities and the environment means that we will have to meet growth in demand and increase resilience to flood, coastal inundation and drought. Achieving net zero carbon will require more efficient use of our available resources and may require us to provide a large volume of additional supply to the energy industry for carbon capture use and storage (CCUS) and the hydrogen economy. With the potential for large investment needs in each sector, cost will become a key driver for decision-makers.

To maintain levels of affordability, measures to further strengthen cross-sector collaboration are necessary, specifically in relation to the development and funding of new infrastructure. Some of this will be relevant for regional strategic issues; others for more local sub-regional or catchment-based issues. Developing single sector (or single company) solutions for meeting future water-related needs in the WRE region, as well as managing the related uncertainties and risks, is unlikely to be cost-effective. A more integrated, holistic, approach is needed.

With this realisation, WRE is developing a number of projects which apply this new approach and blend hydrological, technical, political, economic, environmental, and social aspects. The next section details three such projects: A Fenland Adaptation Strategy; the development of a Regional Natural Capital Plan through Systematic Conservation Planning (SCP); and the implementation of innovative funding models for nature-based solutions through the creation of a Water Fund.

PROJECT CASE STUDY 1: THE FUTURE FENLAND ADAPTATION STRATEGY

An example of where WRE strengthens collaboration between sectors is the Future Fenland Adaptation Strategy. This initiative, which is based on the principles of IWRM, and is applying the MO-RDM methodology, seeks to deliver a long-term solution to the drought, coastal inundation and flooding-related risks which are posed in our Fenland areas by climate change. By coordinating activity and funding in programmes which are traditionally considered to be separate, the overall level of investment which is required can be reduced, and delivery can be made more efficient and the benefits spread more widely. An illustration of the concept is given below.

The Future Fenland adaptation strategy is combining six projects that are existing or planned involving 14 WRE members (public water supply companies, environmental NGOs, Internal Drainage Boards (IDBs), local government, and agriculture).

Sitting at the heart of this overall strategy are two new reservoir systems, one in the south of Lincolnshire and the other somewhere on the Cambridgeshire/Norfolk border (see Figure 5.1 for geography), linked into the network of IDB assets and main rivers and using high and excess flows as potential sources of water for the reservoirs. Combined with potential new barrages on the large river systems surrounding the area (see Figure 5.2), this overall vision has the potential to drive enormous economic, environmental and social benefits which will be felt right across the WRE region.

Key elements of the Future Fenland Adaptation Strategy include:

- New multi-sector reservoirs providing additional water supply resilience for water companies, farmers and the food industry.
- Downstream flood barriers or barrages to protect growth areas in the Fens, enabling key local infrastructure projects such as a rail connection from Wisbech to Cambridge and the dualling of a motorway to move forward.
- Open water channels to provide water storage, biodiversity, navigation and tourism, and further flood risk management benefits.
- The opportunity to collaborate to manage land and water across the Fens in a new and integrated way, seeking to secure the future of the peat landscape given its crucial role in carbon sequestration.

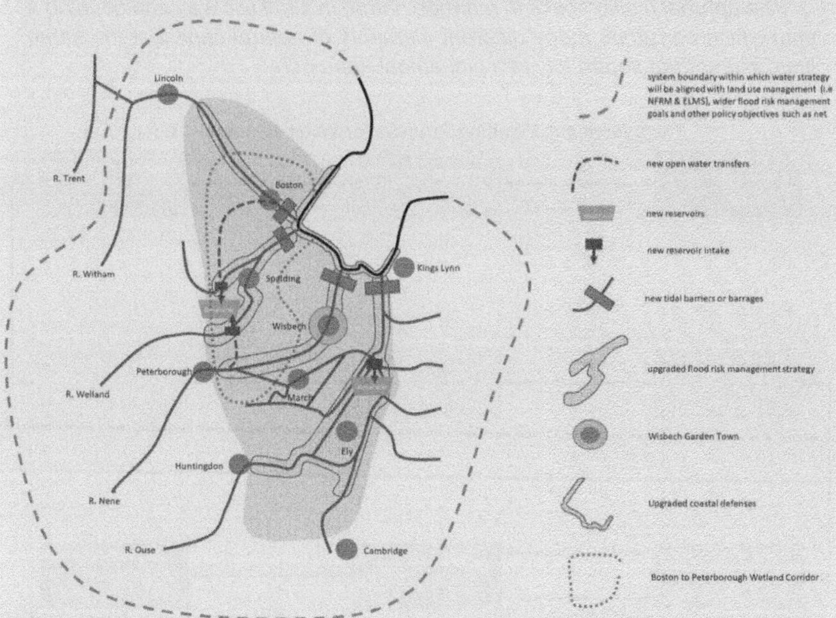

Figure 5.2 Water resources east. *Source*: Price & WRE (2020).

PROJECT CASE STUDY 2: NATURAL CAPITAL PLANNING USING SYSTEMATIC CONSERVATION PLANNING

Developing a Regional Natural Capital Plan Using Systematic Conservation Planning

There is a recognition that WRE needs to find ways to improve and enhance the environment by developing a deep understanding of natural capital in the region to maximise the ecosystem services being delivered. This is being achieved through SCP (see Figure 5.3). WRE is teaming up with Biodiversify and WWF-UK, with financial support from the Coca-Cola Foundation, to develop a natural capital plan for Eastern England. This is the first time the SCP process has been applied in the UK and the first time it has been applied on this scale in the world. The first iteration of the Natural Capital Plan for Eastern England was released in June 2021.

What is Systematic Conservation Planning? by Dr Sam Sinclair, Biodiversify, 2020

Systematic Conservation Planning or SCP is a combination of two things, a social process and a prioritisation analysis. The spatial prioritisation analysis identifies how and where to act to improve natural capital in the most cost-effective manner. This analysis is embedded within a social process which uses an inclusive dialogue to give stakeholders ownership over the plans.

The approach also seeks to manage nature in a holistic way and develop a plan which considers many different elements of natural capital at the same time, rather than separately in a piecemeal approach.

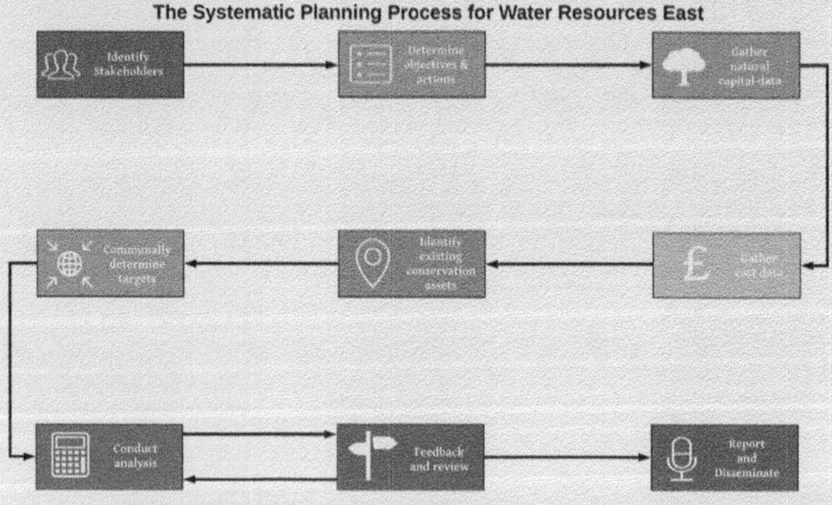

Figure 5.3 Biodiversify. *Source*: Sinclair (2020).

By looking at the bigger picture in this way, the spatial prioritisation analysis looks for synergies; where can you act so that you achieve multiple benefits simultaneously and give the best outcomes for people and nature with the resources available? Because the objectives, actions, and targets are set by the stakeholders the plans also represent the best outcome for everyone; the plans show how people can coordinate across a landscape to achieve common goals as well as having their own needs met.

The purpose of the plan: The main aim of the plan is to develop a shared vision for the restoration of nature across the WRE region. This plan will identify priority areas for different types of natural capital actions across the region. These are not intended to dictate anything or force anyone to do anything they do not want to, instead they just indicate where actions should take place in order to deliver natural capital outcomes as effectively as possible. This is partially about being as effective as possible but also about coordinating action across the area so that everyone is pushing in the same direction.

Stakeholder ownership: We want this plan to be owned by the stakeholders of Eastern England. A plan like this will only be valuable if the people and organisations who live and work across this landscape feel that this reflects their wishes.

We believe that the only way to achieve this is for stakeholders to play a leading role in the development and creation of the plan. To achieve this we will be facilitating a transparent process designed to put stakeholders in the driving seat and give them the power to collectively develop a shared vision for Eastern England.

PROJECT CASE STUDY 3: FINANCING ADAPTIVE, COLLABORATIVE WATER MANGEMENT

What is a Water Fund?

Water Funds are governance and financing mechanisms allowing public and private sectors to work collectively to secure water for their communities (see Figure 5.4). They are used successfully around the world to leverage blended finance streams to ensure coordinated delivery, funding and monitoring of nature-based solutions (NBS) for water security. In 40 locations, across North America, Latin America, Asia and Africa, The Nature Conservancy (TNC) collaborates with partners to set up Water Funds based on science-based plans and innovative tools for representing water management challenges, strong monitoring and mobilisation of diverse funding streams. TNC wish to develop two pilots in Europe; one of these will be in Spain (Murcia) and the other will be Norfolk, England. Being part of the global Water Fund network will provide access to collective experience, accelerating the project, and enable Norfolk to be featured as a global exemplar for water resource management, thereby facilitating access to further financial and human resources (Tremolet & The Nature Conservancy, 2020).

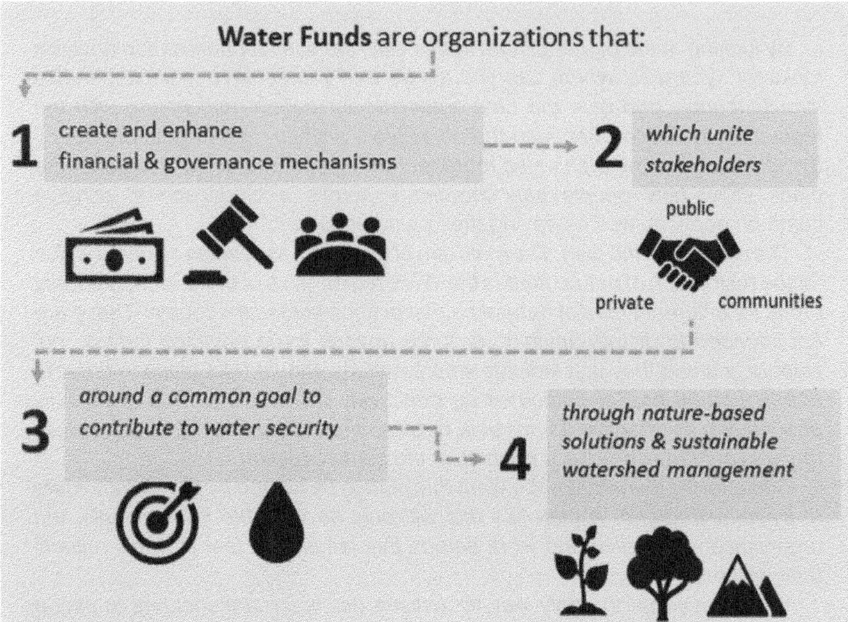

Figure 5.4. The nature conservancy. *Source*: Tremolet S. (2020).

Water Funds and Blended Financing in the County of Norfolk

Our Regional Plan intends to seek out innovative financing options which drive the ability for adaptation, and green returns, and ensure longevity of all projects.

One such financing method utilises the so-called 'blended financing' through the development of a Water Fund. This is a completely new finance approach in Europe, which is being trialed in WRE's North Eastern county of Norfolk.

Norfolk is impacted by a range of water-related issues. These include falling ground water levels, rising abstraction rates, increased housing and business development, and increased flooding events. There is currently no overarching strategy in place to address these issues. This project will develop short-term COVID-19 interventions, and a detailed water management strategy and plan, and establish a partnership structure known as a 'Water Fund' to facilitate delivery in the medium and long term. This project is being delivered by a partnership of Norfolk County Council, Anglian Water, WRE and the international environmental charity, TNC.

The project will create a new multi-stakeholder governance structure which will include representatives from local government, water companies, environmental organisations and the agri-food and energy sectors. This governance structure will be set up in two stages:

(1) A Water Management Board to generate consensus across all local actors for the preparation of a prioritised plan.

(2) A more permanent structure (a Water Fund) to: supervise and coordinate implementation of the plan, monitor results, and enable mobilisation of funding and repayable financing from public and private sources.

Initial work will be detailed analysis of water opportunities and challenges through to the 2050s and beyond (in line with WRE's planning timeline). It will provide a clear understanding of environmental improvements that must be delivered and quantify the impact of current growth predictions and climate change on water availability, water usage and quality. It will be an opportunity to address barriers to COVID-19 recovery and test more ambitious scenarios for future resilience. We will build on existing planning by public water supply company Anglian Water, and WRE, and local plans and economic strategies to focus on Norfolk's specific situation and identify challenges at a more disaggregated geographical scale.

With a detailed understanding of the problem, the project will focus on options to meet the predicted large supply/demand deficit and associated challenges including flood risk, water quality issues and impacts on biodiversity.

Potential options for the plan can be summarised in three main categories:

(1) **Maximising water savings through systematic implementation of demand management measures.** These will include water efficiency measures for new construction and retro-fits to existing buildings, increased efficiency for irrigation through innovation, efficiency measures for other water-using sectors, and working with water companies to ensure effective deployment of smart metering programmes and leakage management. We will identify how new builds could meet ambitious and nationally leading water efficiency targets, and we will continue to go further and faster with regard to water efficiency for all water users, including agriculture.

(2) **Identifying opportunities to maximise the use of nature-based solutions (NBS) to deliver increased water availability and resilience for all sectors and the environment.** NBS are defined by the International Union for Conservation of Nature (IUCN) as actions to protect, sustainably manage and restore natural and modified ecosystems in ways that address societal challenges effectively and adaptively, to provide both human well-being and biodiversity benefits. We will identify opportunities to invest in NBS that can address water security challenges, protect and restore biodiversity, generate carbon benefits (via carbon sequestration or reduced carbon use impacts) and create sustainable green jobs in a wide range of sectors, including construction, agri-food and the visitor economy.

(3) **Identifying the needs for residual grey infrastructure investments.** Grey infrastructure investments such as new multi-sector reservoirs, pipelines and desalination plants are likely to still be needed to address challenges. Considering their development as part of broader, more holistic plans, will potentially reduce their scale, improve integration into the landscape and reduce or eliminate negative impacts.

The project will bring these together in an integrated way to maximise the synergistic benefits and will identify ways of funding the programme in the long term. Opportunities include private sector funding (e.g. from water companies, power companies, developers, the agri-food sector and biodiversity Net Gains credits), public funding (e.g. through flood management schemes, and the new UK Environmental Land Management (ELM) scheme for agriculture), and philanthropic funding.

The success of the above projects and others within the WRE approach rely on having strong collaboration at their core. Without it, trade-offs will not be equitable, consensus and buy-in on projects will not reflect reality, and the future consequences for any water managed could be detrimental.

Already, at the time of writing, WRE operates with members from over 180 organisations and is growing all the time. The next section discusses the power of collaboration for WRE's overall success, how this is applied in practice, and reflects on lessons learnt and recommendations for self-reflection and to pass on to others.

5.6 THE POWER OF COLLABORATION

At the core of WRE's IWRM planning is an active collaboration between all water use sectors. Without the involvement of all, decisions, trade-offs, and future planning are bound to be inequitable.

Natural capital planning through SCP and the development of innovative finance models for nature-based solutions through 'Water Funds' are two ways that our collaborative approaches are being put into action. However, our thinking goes wider than that and involves establishing and nurturing long-term relationships with all water users across multiple levels.

WRE's shared vision is that by working together regionally and nationally across all sectors, we will have a joined-up view of the actions that are needed now, for a sustainable future. Working collaboratively, we will seek to increase the resilience of water supplies, ensure clarity of roles and responsibilities, protect and improve the environment and drive efficiency, providing value for our region.

To put this into a specific context, we are channelling collaborative work in forming the Regional Plan. The overall 'ethos' of WRE's Regional Plan Development will be one of co-creation and engagement, rather than creation and consultation. We are coordinating the outputs of our entire process and the associated working groups via a series of planning conferences for stakeholders in each water use area.

How this works in practice is with WRE and Manchester University teaching and training all members in the process of MO-RDM, how the simulators involved work, what data are used to run the simulators, and creating an understanding of what factors (be that rainfall, population growth, climate change) affect what elements (e.g. water availability for public water supply, resilience to drought and flood events) and therefore determining which water management solutions are best.

It sounds simple, but it comes down to getting everyone together in the same room, learning, having open discussions, and making combined cross-sector decisions about how best to manage water in the future to the benefit of everyone.

This will enable the discussion of a range of proposed solutions for each group, in order to understand challenges and opportunities and to seek common consensus and agreement on the portfolio of options.

The regional training and planning conferences are taking place throughout 2021 and will culminate in a region-wide planning conference where selected options and results will be presented and discussed to go towards the formation of the final Regional Plan due in 2023.

We also hold quarterly meetings with our entire membership and stakeholder base (our strategic advisory group) together with our consultation group (which includes regulators and other government agencies). The purpose of these meetings is to engage our membership in the emerging plan, to identify any concerns or opportunities as early as possible, and to gather feedback and suggestions throughout the process.

In-between planning conferences and strategic advisory group and consultation group meetings, we continue to engage and support on an individual basis with organisations and regularly attend our members' working groups, steering groups and meetings when needed. This creates a 'two-way street' forum to WRE's approach and ensures we're not only 'broadcasting' activities and developments but crucially taking members and stakeholders with us as we co-create the regional plan.

5.7 LESSONS LEARNED

WRE's small core team has seen the programme grow from an Anglian Water idea in 2012, to a project in 2014, and now to a fully independent organisation with, at the time of writing, over 180 members from all water use sectors. The opportunities that we have recognised and the lessons that we have learned have been vast – there have of course been obstacles and difficult moments – but these have all helped to make WRE more resilient, improve our experiences, and share our successes as well as learn from our mistakes.

Our core team and some of WRE's longest standing contributors reflect on some of those lessons as well as recommendations to others who may want to apply aspects of our approach to integrated water management and planning.

Dr Geoff Darch, Water Resource Strategy Manager, Anglian Water

'Water resources have never been so important. We provide more than a billion litres of clean, wholesome water a day to our customers in the driest part of the UK. We are already seeing water resources under increasing pressure and with future climate change balancing supply and demand will become an even greater challenge. We believe the best outcomes will be facilitated by working with all stakeholders on multi-sector solutions that can deliver a balanced outcome for the environment, society and the economy.'

Henry Cator, Independent Chairman, Water Resources East

'As with any new strategy you need to bring people together to make decisions which they can believe in. In order to achieve that cohesive partnership one needs to build trust and mutual respect between all concerned. It is amazing what will be achieved if folk work together using positivity to overcome any hurdles. Communication is key at every level and each stage.'

Martin Collison, Engagement and Policy Advisor, Water Resources East

'The challenges on water management have built up over many years and so it will take commitment and investment over many years to deliver the long-term water security we need.

We should not expect the responsibility for addressing water management challenges to always fall to others, be it water companies, drainage boards or the Environment Agency, every household, business and community has a key role to play. Everyone can help through the actions we take to conserve water or to manage our impact on flooding through using sustainable drainage systems or managing our land or gardens, so they reduce runoff.'

Nancy Smith & Rachel Dyson, Communications and Engagement, Water Resources East

Nancy Smith: 'Building long lasting, trusting relationships has no deadline and your strongest stakeholders are often your loudest critics.

If you can, have a social media presence for your organisation or project. A pleasure as much as it is a pain, it extends reach, spreads news and research, and could lead you to all sorts of people!'

Rachel Dyson: 'Collaborative, cross-sector, integrated water management, while a seemingly obvious way to manage water, is a radically different approach to how water has historically been managed, particularly in the UK. Therefore, it should never be underestimated the amount of time and effort required to understand your stakeholders' views and to build their understanding of the various compromises and trade-offs required so that ultimately all water users' needs and those of the environment are balanced fairly.'

Dr Robin Price, Managing Director, Water Resources East

'WRE operates in a part of the UK with very high economic and environmental ambition. Stakeholders such as those planning for additional housing or business development or looking to boost agriculture and food production have often sat on opposite sides of the table from those who wish to restore and enhance the environment. We have found that our structure, approach and unique set of tools have helped to start to bring people together to plan for the future – recognising that actually, everyone wants the place that they live to thrive economically, but that this can't be achieved fully unless the environment thrives. Integrated, holistic water management is a wonderful golden thread which runs through this conversation.'

Dr Steve Moncaster, Technical Director, Water Resources East

'Main lesson I've learned from the WRE process is that effective collaboration takes a lot of time & resources and that the key to success is openness & transparency. When we talk to people in the States with experience of this kind of stuff they same the same. According to them, you need to build allies & alliances and once you've done this you need to make sure that these are maintained – potentially over very long time periods (20-years + needed to deliver the really big schemes).

In short WRE is as much a social enterprise as it is a technical one and we're here for the long-haul!'

Peter Simpson, Chief Executive Officer, Anglian Water

'For me 'WRE' started when we (Anglian Water) realised that our approach to water planning needed to change in response to the challenges of a growing population and the impact of climate change. In short it needed to move from a fixed, single sector, view of the future to a multisector scenarios based one. And to an approach where all of the users of water have a stake with damaging the environment in a drought no longer seen as a fall-back option.

What have I learnt:

- As always, the power of a clearly communicated vision.
- The benefits of co-creating the organisation, governance and consultation.
- Not leaving 'strategy' to someone outside the room but focusing instead on ways of building ownership of it – ultimately if you were involved in building and it doesn't work then guess what you'll sort it.
- Picking the team carefully and having passionate enthusiasts up front as you must win hearts as well as minds.
- Work with the grain – where there are projects which you can align with, do, as not everything has to come from you.
- Don't forget the basics – there has to be a Water Resource Management Plan in the end!'

Jean Spencer, Chair of the National Planning Framework

'WRE is a model in developing a multi-sector collaboration for water resources and the environment. It has taken many years to reach the position where all sectors have an equal voice in developing the plans for the region. The next step will be to the development of mechanisms for all sectors to play their part in taking actions, as well as funding investment, to improve the environment for all.'

5.8 CONCLUSION

WRE as an organisation is still in its infancy but we believe we are well positioned through our approach, our inclusive governance structure and our technical toolkit to truly implement a new and much more joined up way of working on how water resources are planned and managed in Eastern England and beyond, given the threats to water supply. WRE hopes to demonstrate leadership, provide a clear and strong evidence base and become an international exemplar.

This comes too with the understanding that all water realities are contextual; with no one approach acting as a panacea for any one challenge. Nevertheless, we hope this chapter has shown how the methodologies and approaches of WRE are working to ensure the sustainability and reliability of water in our corner of the world.

REFERENCES

Moncaster S. (2020). Water Resources East Method Statement for the WRE Regional Water Resource Plan. Water Resources East. Available at https://wre.org.uk/wp-content/uploads/2020/08/WRE-Method-Statement-V5-FINAL.pdf.pagespeed.ce.zdXLTOnVve.pdf (accessed 1 November 2020).

Price R. (2020). Water Resources East: Initial Water Resource Position Statement. Water Resources East. Available at https://wre.org.uk/wp-content/uploads/2020/04/WRE-Initial-statement-of-resource-need-FINAL.pdf (accessed 1 November 2020).

Sinclair S. (2020). 'Developing a Natural Capital Plan for Eastern England' [PowerPoint Presentation] Biodiversify.

The Environment Agency (2020). Meeting our Future Water Needs: A National Framework for Water Resources. The Government of Great Britain and Northern Ireland. London. Available at https://www.gov.uk/government/publications/meeting-our-future-water-needs-a-national-framework-for-water-resources (accessed 1 November 2020).

The Government of Great Britain and Northern Ireland (2017). Industrial Strategy: Building A Britain fit for the Future. Department for Business, Energy and Industrial Strategy. London. Available at https://www.gov.uk/government/publications/industrial-strategy-building-a-britain-fit-for-the-future (accessed 13 April 2021).

The Government of Great Britain and Northern Ireland (2018). A Green Future: Our 25 Year Plan to Improve the Environment. The Environment Agency. London. Available at https://www.gov.uk/government/publications/25-year-environment-plan (accessed 1 November 2020).

The National Farmers' Union (2018). Learning Lessons From the 2018 Agricultural Drought. The National Farmers' Union. Newmarket. Available at https://www.nfuonline.com/nfu-online/science-and-environment/climate-change/221-1118-leasons-learnt-drought-2018-final/ (accessed 26 January 2021).

Tremolet S. (2020). 'Mobilizing investments in nature-based solutions for water security: A Water Fund for Norfolk' [PowerPoint Presentation] The Nature Conservancy.

Chapter 6

Implementing integrated water resources management locally in rural catchments: Lessons from eastern Sudan

Khaled Mokhtar[1] and St John Day[2]

[1]*United Kingdom Foreign, Commonwealth and Development Office Sudan*
[2]*Principal Consultant, IOD PARC, Edinburgh, United Kingdom*

ABSTRACT

Sudan is a vulnerable and challenging environment as a result of its climate, hydrology, and hydrogeology. Other entrenched human factors, such as authoritarian rule, limited historical investment in rural water services and the gradual decline of national institutions make it particularly difficult. This has manifested itself today into low levels of water supply coverage particularly amongst rural communities. Trust between rural communities in Kassala and government institutions has also declined for those left behind in rural hinterlands. Providing sustainable and resilient water services in rural Sudan is difficult work, not least because of high rainfall variability, inadequate infrastructure and the lack of continuous external support to communities when problems arise. This paper describes efforts to strengthen links between water resources management and WASH, and the challenges faced when national institutions responsible for water resources and water supply are weak. It documents recent efforts to ensure water supply services can provide water year round and increase collaboration between rural communities and mandated government authorities. It is intended to be read by government personnel, non-governmental organisations and other staff that are directly involved in implementing integrated water resource management programmes in complex environments.

Keywords: water resources management, arid environments, community-based.

Over the past decade or so there has been growing interest in water resources management in humanitarian and development programming in Sudan. External donors (such as UK Aid), national institutions and multilateral agencies are increasingly focussed on sound stewardship of water (and land) resources in the knowledge that parts of Sudan (such as Darfur and Kassala) are prone to increased water scarcity and drought. This focus has been influenced by a number of reports and articles that have argued for better integration between water resources management and WASH in an effort to support peace and recovery (see Tearfund, 2007).

Sudan is a particularly challenging environment. One of the key requirements is to determine how sound stewardship of water resources and greater community resilience can be achieved in such difficult working contexts. The most obvious aspect, perhaps, is the arid environment and the remote nature of rural villages with their multi-dimensional poverty, as a result of limited access, weak infrastructure and poor communications. However, the challenges posed by human-induced factors are also problematic. This includes: armed conflict and displacement; authoritarian government rule and disruption to traditional governance arrangements; limited investment in development programming over many decades, which hinders access to water supply and sanitation services; the decline of national institutions; and subsequent widespread mistrust between rural communities and government. Sudan's challenges go way beyond the natural environment and the unfavourable political and institutional arena means the commitment and capacity to deliver resilient and sustainable services are absent.

The recognition of the multiple uses of water in Sudan, for use at home as well as for agro-pastoralism, immediately places the spotlight on the importance of water resources management. Perry (2008) provides an informative summary of the five key elements required to ensure sound stewardship of water resources. The first requirement is the availability and quantity of water resources needs to be known and understood with a reasonable degree of accuracy. This can be achieved through local hydrometric monitoring of rainfall, surface water and groundwater; and based on past experiences of how water resources respond to rainfall. The second aspect is that different and competing water users have arrangements in place to share water at many levels – sub-catchment, catchment and transboundary. This requires water usage to be prioritised and a bargaining process between upstream and downstream users so water is shared equitably. The third aspect requires any understandings over water allocation and usage to be bounded by agreed rules. This can take the form of operating principles at local or community-level, byelaws at catchment level or more formal water laws at national and transboundary levels. The fourth aspect – that institutional roles and responsibilities are clearly defined – is common in policy and practice. Multiple institutions at many levels need to be involved in managing water resources and this includes community-based groups. The fifth aspect is concerned with the provision of physical infrastructure to harness, store and supply water in an efficient manner. In order to apply these five elements successfully the wider systems or networks of policies, laws, people, institutions, infrastructure and finance need to be performing well. If they are absent then interventions will need to be adapted and tailored carefully to the local context. This is an important consideration because state institutions in Eastern Sudan are perceived as having limited capability and resources.

Over the past three decades, the concept of integrated water resources management has been the dominant paradigm in international and national water policy for managing water and land resources. It is a broad and all-encompassing approach and inevitably it needs to be unpacked so it can be applied appropriately – according to the challenging nature of the local context. This is particularly important in a fragile and conflict affected setting, like Sudan, where the enabling conditions for integrated water resources management are absent. For example, the first aspect of Perry approach which we have highlighted, focuses on assessing water resource availability. In real terms local hydrometric monitoring networks are often absent in Sudan and capacity within mandated institutions to collect, analyse, process and publish hydrometric data is weak (in the sense of limited skills, knowledge and resources). This means it is difficult to assess water resources availability on a continuous basis. Furthermore, physical water storage and recharge infrastructure is often absent or in a state of disrepair. The case study in this chapter highlights efforts by Plan International to introduce integrated water resources management in challenging contexts. It focuses on experiences from Kassala, eastern Sudan.

6.1 CONTEXT ANALYSIS

Sudan is the third largest country in Africa, with an estimated area of 1.88 million km^2. In the far north-east, Sudan is bordered by the Red Sea and it shares common borders with seven other countries, namely: Eritrea and Ethiopia in the east, South Sudan to the south, Central African Republic and Chad in the west, Libya in the far northwest, and Egypt in the north. Regional insecurity and the porous nature of Sudan's borders has had a significant historical impact on its internal security. One devastating factor, for example, has been the influx of firearms from Libya and the resultant localised conflict between different ethnic groups – fuelled by the previous regime in Khartoum.

Kassala state (Figure 6.1) is one of 18 states in Sudan and it is divided into 11 localities. Rainfall in Kassala, typically occurs during the months of June to mid–September and average annual rainfall is low. For example, in the Hamesh Koreib catchment average annual precipitation is <200 mm and accompanied by

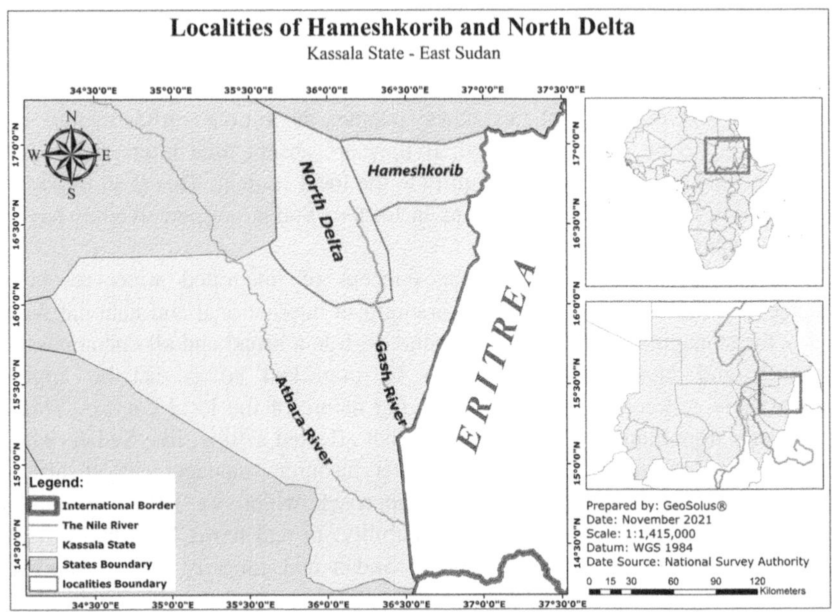

Figure 6.1 Kassala state and intervention areas.

high evapotranspiration rates (around 90%) with corresponding low groundwater recharge.

Pastoralism was once the dominant livelihood in Kassala state, but this has been gradually reduced due to successive droughts, causing the loss of animals and in some cases entire herds. The expansion of agro-pastoralism and farming along transhumance routes has also limited access to grazing land and reduced pastoralist activities. It is estimated that the nomad population has reduced from 50% of the population in the 1950s to around 10% in the 1990s. However, people still do not fall into the neat categories of urban and rural dwellers. In these remote rural areas, some pastoral and agro-pastoral communities still move locations according to season and pasture availability.

The Gash river is the dominant water supply source for Kassala state. It originates in central Eritrea and is transboundary, flowing into Sudan from the east. The river is also seasonal and the flood waters typically retreat in August, allowing the progressive sowing and cultivation of sorghum during the months of December and January. This is a vital period for rural communities for food production and marketing products ahead of the extensive dry season, which lasts from late September to June.

Both the Hamesh Koreib and North Delta catchment areas and the people within are characterised as suffering from: widespread food insecurity, high rates of malnutrition, poor access to health services, limited access to water and sanitation services and limited access to education. There is also high social inequality between men and women, and many girls and women are restricted from accessing basic education and income generation services. This is particularly true in Hamesh Koreib, which is a more conservative and traditionalist locality.

While resilience challenges are multi-faceted, the real water security concern for rural populations in Hamesh Koreib and North Delta is the risk of consecutive low rainfall years leading to declining groundwater and reduced water access. Historically, this has resulted in increased water disputes between agro-pastoralists and nomadic pastoralists, and an unwillingness to share water resources between upstream and downstream communities.

Across Kassala state, the disparity in access to water services between urban and rural communities is also well documented and borne out by the resentment of communities towards state institutions that are mandated to supply rural areas. Population growth rates are increasing in both urban and rural areas and this places increased pressure on water resources. Despite the steady population increases in rural areas, often the younger and more mobile members of society migrate to towns or cities in search of employment and new opportunities. This has created further challenges as the elderly and more vulnerable members of society remain left behind in rural hinterlands, forced to operate and manage water supply services. Marginalised groups often suffer from extended periods of water scarcity, where water demand exceeds available supply.

Specific problems related to water resources management in rural Kassala include: high climatic variability leading to drought or risk of seasonal flooding, water source pollution, growing water demands and climate change. Limited hydrometric monitoring, coupled with inadequate capacity within state institutions that are mandated to monitor and manage water resources is a further pressing concern. This means any infrastructure programming is often not well targeted. This is an indication as to how the policy environment for integrated water resources management approaches needs to be strengthened in order to sustain programme interventions by international non-government agencies (INGOs).

Anecdotal information recorded by Plan International suggests that people in Hamesh Koreib and North Delta are observing increased climatic variability and greater seasonal variations in groundwater levels. For example, boreholes and wells are 'perceived' to becoming seasonal – where previously they provided water the year round. Seasonal water fluctuations, whether due to increased demand or climate change, has a direct and adverse impact on domestic and multi-purpose water usage. Limited recharge capacity and reduced access to water has led to intense competition over fertile land and reliable water sources. This threat has also perceivably increased following the decline of traditional governance systems in some localities caused by political interference. Consequently, planned interventions by INGOs need to be aware of these growing environmental, political and societal pressures in order to ensure their interventions are appropriate to the local, national and regional context. An important consideration in such challenging environments is that water resources management should not be undertaken for its own sake and the ultimate aim is to ensure more sustainable and resilient water supplies for all users. This is no easy task given the human factors that hinder sound development practices.

6.2 INSTITUTIONAL CHALLENGES

Much needed efforts to address water security problems have been hindered in the recent past as a result of authoritarian rule from Khartoum. One example, often highlighted in conversation with water sector professionals, is the deterioration of institutional capacity when the Rural Water Authority (RWA) that was dissolved in 1996. This issue is discussed here. Previously, the RWA was managed centrally in Khartoum with the overarching mandate to regulate, plan, raise funds, and maintain water supply services for rural villages. Historically, the policy intent of the central government was to give state governments more power in terms of managing their own resources which included the provision of urban and rural water supply services. However, one of the main challenges that emerged since 1996 was the increasingly limited and inadequate decentralised financial and technical support received from central government. Consequently, the responsibilities of managing water successfully were handed-over to state governments, but control over financial and human resources remained with the federal government in Khartoum. To highlight the flawed financial policy, taxes and tariffs collected at state level were transferred back to central government, to then be 'redistributed' again based on the federal government's priorities. This, perhaps unsurprisingly, led to increasingly late and inadequate financial disbursements to state institutions as money was retained by the central government. Furthermore, the priorities of an authoritarian government, desperate to remain in power, were rarely aligned to people's development needs at the state level. This led to increased poverty, weakened institutional capacities and the loss of trust between government and its citizens. For example, historical interventions reportedly operated for five years before the state government, participated in providing any meaningful financial support for the construction, rehabilitation, or maintenance to rural water supply services. External technical support has continued to be limited, and no local staff from the new State Water Corporation (SWC) were assigned to assist rural communities in the Hamesh Koreib catchment area. This led to growing inequity between urban and rural areas and increased mistrust between government agencies and people living in remote rural hinterlands.

Indeed, the lack of engagement between state agencies and rural populations has led to widespread resentment and mistrust, which continues to this day. Rural households and communities, when discussing their development priorities with Plan International, initially refused to include government representatives in the meetings. In some extreme situations, government involvement is seen as a hindrance to building sustainable and resilient water supply services. This reflects the low level of trust that exists. A commonly held view by some water sector professionals in Sudan is that the replacement of the federal RWA with the SWC has not achieved the desired impact. The SWC now predominantly focuses on urban and peri-urban populations with minor consideration for rural areas. Indeed, INGOs and NGOs are expected to fill this service delivery gap, however, such interventions are often relatively short lived and unable to take a long-term sustainable water approach. This means once funding declines or programmes focus elsewhere, there is justifiable concern as to how development progress will be sustained. Consequently, rural populations feel left behind and there has been a continuous decline in access to basic water and sanitation services. Ongoing external support for community-based maintenance (CBM) of rural water supplies has also declined.

For some donor partners, there are also other institutional aspects in Sudan that are controversial. The Sudanese Government's Water and Environmental Sanitation Department (WES) is part of the Public Water Corporation, which in turn is part of the Federal Ministry of Irrigation and Water Resources. They play a prominent role in rural water supply implementation and monitoring as well as humanitarian assistance. However, there is the very real risk that institutional roles and responsibilities are duplicated as WES works at federal level and SWC operates at state level. WES undertakes important work but some external donors believe there is a necessity to reform institutional arrangements for integrated water resources management and rural water supply, not least to improve accountability to rural populations.

6.3 WAR AND CONFLICT

In 2003, a devastating civil war erupted in the eastern region of Sudan. Kassala was one of the major battle fields for the rebel forces and the government. Conflict and violence destroyed much water supply infrastructure and punctuated any development progress. The impact of the civil war can still be felt today with unexploded landmines and ordinance in many places that hinders movement and access across vast rural areas.

In 2006, a Comprehensive Peace Agreement (CPA) was signed between federal government and rebel groups. It included an ambitious plan to rehabilitate and develop physical water supply infrastructure for both rural and urban people. However, these ambitions have not come to fruition. Yet again development focus shifted from rural to urban populations with no corresponding improvement in the government water management structure or updated mandate for SWC.

In 2011, largely due to the secession of South Sudan and the corresponding economic instability that followed, the Government of Sudan's expenditure on rural water supply virtually collapsed. Today this finance gap is partially filled by external donors who implement projects led by UN agencies and INGOs.

However, finance (capital expenditure) is limited and inadequate and government makes negligible contributions for ongoing recurrent finance. This means the responsibility for all post construction finance falls on communities.

As a result of these factors and the absence of a clear strategic vision to manage water resources in Kassala state, rural water supply performance remains weak. The remainder of this paper describes efforts to pilot integrated water resources management in the Hamesh Koreib and North Delta catchment areas and to share lessons. It concludes by sharing thoughts on future work and opportunities following the Sudanese revolution in 2019.

6.4 AQUA FOR SUDAN

To help address some of the water security challenges mentioned above, UK Aid supported a multi-agency programme that promotes integrated water resources management in Sudan. This included Red Sea state, Kassala and Gadarif states, as well as northern, southern and western Darfur. The main objective of this programme was to pilot water resource management interventions as part of conventional water, sanitation and hygiene (WASH) programming. Lessons learnt were documented and shared across the consortium.

WASH programmes are primarily focused on delivering water supply and sanitation services to rural communities and bringing about changes in hygiene behaviours. CBM models are widely promoted in Sudan, because they are considered to be the most viable option for keeping rural water services functioning. However, the CBM approach is not without its detractors because service breakdowns often exceed local capacities, which results in people returning to more distant and unprotected water sources. Nevertheless, in fragile and conflict affected states there are strong reasons for pursuing with this approach. Three in particular stand out. First in Kassala state there are many remote communities that are far removed from state institutions let alone private providers who can support rural water supplies. The external support arrangements that are desired to keep water supplies working are virtually absent. Second, rural communities often have great strengths and resilience. They live in harsh environments and face continuous hardship and threats, such as drought. Communities use water for both domestic and productive use to support their livelihoods and often they have developed their own informal water management arrangements and practices that should be better understood and supported where appropriate. Third, service providers can do much more to ensure rural water services implemented are relevant to the local context and implemented effectively in order to minimise the potential for service disruption.

To manage water resources effectively requires a different set of skills, but there are still many actions that can be performed by local community-based institutions. For example, farmers can be active participants in monitoring local water resources and communities need to be active participants in bargaining over water allocation in catchment areas. Rules, or operating principles, for water usage and management need to be grounded in practical realities by those that can police usage on a day-to-day basis. Institutions at many levels need to be active participants in managing water resources, but they should also be subservient to 'lower' level institutions. For the reasons stated related to CBM, rural communities also need to be active participants in managing and maintaining physical water infrastructure (such as recharge dams and wells).

This chapter covers experiences in two sub-catchments in Kassala state (namely Hamesh Koreib and North Delta) that were implemented by Plan International alongside local community-based institutions and government agencies. Both sub-catchments are located within the boundary of Al-Gash River. The Al-Gash is a seasonal river that typically flows between July and October, however on rare occasions river flows may start earlier or run later. The river serves as the main source of groundwater recharge for many adjacent wells and boreholes and is the main source of water for Kassala city and other towns in Kassala state.

The Al-Gash catchment is transboundary, and its water resources are shared between Sudan, Ethiopia and Eritrea. However, there is little transboundary cooperation between respective state institutions for processing and sharing river flow data. The size of the catchment is 31,000 km^2 and is characterised by low average annual rainfall, and high evapotranspiration. The catchment area also includes vast areas of fertile soil, which is used to cultivate vegetables and fruits for Kassala city.

North Delta sub-catchment is located in the delta of the Al-Gash River. At peak flood the Al-Gash water floods adjacent arable land. People in North Delta rely on seasonal agriculture as their main source of livelihood. Animal husbandry is practiced and includes both cows and camels. Water sources within North Delta rely on shallow groundwater storage that are fed through a series of canals constructed as part of the Al-Gash Agriculture Scheme. Geological features of the North Delta catchment do not allow for deep groundwater recharge.

The more remote Hamesh Koreib locality sits within a sub-catchment of Al-Gash River basin, but beyond the upper reaches of the river. Agro-pastoralists in this area are reliant on natural resources and typically grow basic food stuffs like sorghum or practice animal husbandry, mainly camels. These livelihoods are dependent on groundwater for their survival and people are often forced to migrate during periods of seasonal water scarcity.

The most recent water security threat associated with the Al-Gash river basin is as a result of the civil unrest in the transboundary countries and frequent border tensions that arise. Human displacement and migration from Eritrea and Ethiopia into Sudan places significant pressure on the water and land resources of host communities. This has led to a situation whereby there is the perception that water resources are being exploited beyond their natural limits and beyond the capability of government institutions to effectively monitor and manage the resource. This has also increased competition over natural resources (fertile land, wood and water) between different ethnic groups. Furthermore, the decline of traditional governance arrangements as a result of deliberate political interference and displacement, coupled with entrenched mistrust between rural populations and government authorities, makes the enabling environment for water resources management extremely challenging.

6.5 APPROACHES FOLLOWED

Given these multiple problems, and the potential solutions offered by integrating water resources management and WASH approaches, the consortium decided to conduct pilot activities across Sudan. The intention was to strengthen the 'water' component in WASH by improving water availability year-round. The interventions were also focused at sub-catchment scale rather than isolated villages within the catchment area, to build a mentality that ensures equitable and sustainable use across communities sharing the same catchment resource. The process involved forming a multi-agency consortium so lessons could be shared widely in order to influence national policy and practice, and attract further investment. The consortium was overseen by ZOA but partner agencies (Practical Action, Plan International, International Aid Services, SOS Sahel and Islamic Relief) were encouraged to follow a systematic approach that could be adapted to the local context. This would help to ensure interventions were relevant to the needs of rural communities. Indeed, a community-based approach was pursued based on evidence showing that in conflict affected areas basic services are better managed at the community level. One of the intentions of the Aqua for Sudan project was to determine whether this could be achieved using an integrated water resources management approach.

The approach adopted by the consortium is set out below with justification following:

- Map and document key characteristics within target catchment areas.

  This was undertaken to identify the key hydrological, hydrogeological, meteorological parameters in order to assess the local water balance, as well as identifying relevant socio-economic and ethnic characteristics of communities living within the catchment area that drive choices around water consumption and management.
- Identify the specific water security problems in all target catchments

  To ensure interventions were relevant and effective, Plan International explored the needs of all water users across the catchment areas. The pressures are multiple, but clearly growing demands on dry season water availability is one of the most pressing concerns.
- Establish catchment level management committees

  This was needed to start building capacity of, and trust between, rural communities sharing water resources over a catchment area. They were then used to facilitate planning of programme interventions and build linkages with government representatives.

- Launch a Water Resources Committee in Kassala state

    In building stronger links between rural communities and government agencies, a new committee was established to foster dialogue and promote sound stewardship of water resources.

- To construct and rehabilitate physical water supply infrastructure

    Based on the water resource diagnostics and subsequent catchment committee preferences, physical water supply infrastructure was built to improve storage and supply of water, including simple groundwater recharge structures to extend the availability of groundwater into the dry season.

- To promote corresponding improvements in sanitation and hygiene promotion

    In order to maximise health benefits, efforts were made to minimise open defecation, promote sound hygiene practices, safe water handling, storage and consumption.

    To document lessons learnt, share findings and influence national policy-making.

    Despite not having the necessary human, equipment and financial resources to re-establish water resources management at multiple levels, this project did want to showcase what was possible in improving water security and resilience for remote and vulnerable communities in rural Sudan. It is hoped this will serve to lay the foundations for similar work in the future.

6.6 PROGRAMME ACHIEVEMENTS
6.6.1 Forming of catchment management committees

As a starting point integrated water resources management interventions must make a difference to people's lives. The process followed was highly participatory to ensure interventions are relevant and there is real demand because it requires significant commitment to behavioural change by host communities. The approach followed was to engage with key stakeholders and community representatives in both Hamesh Koreib and North Delta catchments from the outset. Meetings were conducted in local dialects and focused on understanding what tasks rural populations could realistically perform. This approach paved the way to mobilise communities towards the establishment of catchment management committees with sound knowledge and concrete understanding of the planned intervention benefits. Committee representatives were selected by the communities themselves and member profiles required the following attributes: possessing negotiation and inter-personnel skills; awareness of multi-water usage;

knowledge of basic water management concepts; ability to devote time to water security issues and trusted by other community members. The catchment committee also included representation from marginalised groups (disabled) and local government. In Hamesh Koreib, a separate women's water management committee was established due to the conservative nature of ethnic groups that maintained male/female segregation practices. In North Delta, catchment management committees consisted of 30% female representation, although active and meaningful participation of those women was recognised as a real challenge.

It can be easy to conclude that once catchment committees are formed they function well. However, two requirements in particular stand out. The first is that integrated water resources management should not be undertaken for its own sake and committee members need to see the benefits and relevance of programme interventions. Some interventions are clearly more popular than others. Infrastructure (such as recharge dams) was a priority in Kassala state because people felt infrastructure improvements were necessary to improve water availability and access. Committee members are living and working in difficult environments and development needs to lead to positive change, otherwise their participation will decline. Infrastructure interventions need to be implemented professionally to ensure impact, but they also need to be accompanied by a process of evaluation and learning so improvements in water availability can be quantified. This can serve to build up evidence of what works and reassure rural communities that interventions are relevant to their water security priorities. The second is that state institutions who are mandated to manage water resources are starting from a low base and it takes time to develop their capabilities. Where government funding and support is inadequate, external funders need to collaborate to ensure these institutions can provide a basic external service to catchment committees. An example of the support tasks to be performed is provided in Table 6.1. This is suggested as a checklist of what external support to catchment management committees could look like.

When government institutions provide effective external support there is extensive evidence of community-based approaches working well. However, years of mistrust and limited engagement cannot be overcome rapidly. There must be faith that government actions will have a tangible beneficial impact on communities' needs and they are willing and able to provide development assistance routinely. Trust will not be achieved if government interventions are lacking or not relevant to people's needs.

6.6.2 Provision of physical infrastructure

Following discussions with community representatives, it was also evident that project interventions should focus on increased access and availability to water. Indeed, physical infrastructure (recharge dams) were seen as being the best way to stimulate change and improve local water security and resilience.

Table 6.1 Responsibilities of catchment management committees and state institutions.

	Responsibilities of Catchment Management Committees	Responsibilities of State Institutions
Assessing water availability	• Monitor and record daily rainfall • Monitor and record groundwater levels • Together with state institutions thoroughly discuss the results and implications for dry season	• Approve and install monitoring instrumentation • Support data processing, analysis and publication • Discuss interpretation of results with catchment management committees
Water allocation	• Ensure water is used equitably within the catchment. • Prioritise water usage. • Engage with large-scale abstractors.	• Supporting for engagement between upstream and downstream water users. • Ensure a bargaining process is adhered to. • Thoroughly discuss water availability based on available monitoring data.
Water laws	• Establish rules or operating principles for water usage. • Established systems for dispute resolution. • Enforcing graduated sanctions for rule violations.	• Support the development of local governance arrangements. • Ensure national water laws are published and implemented. • Ensure external support for enforcement.
Institutional roles	• Managing water locally. • Policing day-to-day water usage. • Insist that Catchment Management Committees formerly recognised in national policy and law.	• Ensure recognition of roles and responsibilities at national, state and community levels. • Provide continuous capacity building support to Catchment Management Committees.
Water infrastructure	• Active participants in technology choice and siting infrastructure • Provide labour and material during construction • Financial contribution (if possible) • Minor operation and maintenance duties	• Support for design, construction • Support for procuring equipment. • Ensure high-quality supervision during construction • Hand over infrastructure to community.

In order to learn 'what works' the programme considered a number of potential groundwater recharge mechanisms. This included: rehabilitation of existing small dams; building check dams or wing dams to reduce sediment build-up; and, construction of new sand dams and sub-surface dams. When determining what is the most suitable intervention, the programme took into consideration the available expertise to build these structures effectively, and the ease with which they can be maintained by the community, given the prospect of external support from government was extremely limited.

As part of the initial planning it was identified that the construction of new small dams in remote rural Kassala was overly complex. Both in terms of design, ensuring construction quality and the ease with which construction materials could be transported to site. The technical skills required were considered to be beyond the capacity of the few local contractors prepared to work in Hamesh Koreib and North Delta. The dams would be subject to high sediment loads and maintenance requirements. The programme also critically questioned the impact of small dams in terms of restricting wadi flows for downstream communities. It was felt this may not constitute good change as the risk of local water disputes was high. However, during this process two existing dams were identified for rehabilitation. An impact assessment on nearby groundwater wells indicated that, as a result of the groundwater recharge infrastructure constructed, the seasonal range or fluctuation of groundwater had reduced by around 0.6−0.8 m, which provided an indication of increased water availability. This rehabilitation work was complimented by the addition of check dams to reduce sedimentation load.

Other examples of programme interventions included the construction of sand dams, artificial recharge basins, plus improved access to hand dug wells and water distribution yards. Due to the remoteness and difficult nature of the working contexts, many discussions with community representatives were held to ensure the interventions were relevant to the local context. Manual well digging and the construction of sand dams and recharge mechanisms made them particularly relevant to shallow unconsolidated geological formations. Private water yard operations were also deemed a good fit for a context where the population has a large number of livestock and people aspire to access fertile grazing land and water. However, it is important to ensure the selected or appointed water yard operator is regulated by the community through a set of operating principles and a willingness to ensure service levels are both agreed upon and maintained.

6.6.3 Documenting and sharing learning

Another important process followed was to document and share learning. This was achieved by documenting case studies and engaging with a wide range of stakeholders at local and national levels. As part of Plan International's programme work, it was decided that integrated water resources management programmes need to focus attention on five aspects: first, to determine the critical water security problem that people experience and ensure subsequent interventions are relevant. Second, to ensure interventions are both relevant and effective in order to bring about improvements in people's water security. This means programmes need to focus attention on physical infrastructure, but also determining impact with regards to improved access, availability, quality and reliability of water. Third, intuitively integrated water resources management programmes should generate learning concerning water resources. For example: how quickly does groundwater respond to rainfall? What is the seasonal range of water level variation? How fast does groundwater recede? How does the catchment area respond to rainfall? How can communities and local authorities monitor water resources? How can raw data be transferred to analysed information? What are the main water security risks to address? With hindsight the measurement of water resource impacts around new infrastructure (such as check dams) should have been prioritised sooner to demonstrate impact. There are clearly dangers in trying to do too much through a short programme intervention but a priority must be to generate learning on water resources and local hydrology. The fourth fundamental requirement is to document what happens when problems arise, such as water disputes or the breakdown of water supply infrastructure. What actions can catchment committees realistically undertake and what rapid external support can mandated authorities provide? These inevitable problems need to be considered with hard-headed reality. The fifth aspect concerns wider systems strengthening work. Inevitably it will take years, even decades, to build effective local, national and transboundary systems for monitoring and managing water resources. Three short-term fundamentals for systems strengthening are as follows: first, to support a decentralised approach for water resources management so local authorities can better fulfil their mandate and ensure accountability to rural communities. This is particularly important given Sudan's recent history. Second, to establish increased and assured budget locations for capital and recurrent finance so institutions can support catchment management committees; and third, to focus on institutions performing essential functions routinely. Later in this chapter we set out what some of these functions are.

6.6.4 Preparation of localised water security plans

Given the considerable water security problems in Hamesh Koreib and North Delta catchment areas, collaborative partnerships also needed to be established between catchment committees and key government agencies.

The Kassala Water Resources Commission (WRC) was established in 2019, and members included representatives from both catchment committees, Groundwater and Wadis Department, SWC, NGOs and INGOs. In order to ensure the commission plays a useful role it needs to be committed to address technical and social matters by undertaking its own studies and interventions locally. However, the commission must also consider how it can advocate for more technical, financial and human resources from external donors and the state.

Having brought stakeholders together our programme experiences show it is vital to build trust and demonstrate that the partnership can resolve water security problems that people experience. The WRC needs to work directly with both catchment committees to stimulate a good change. This requires time and effort from all parties. Specific problems need to be unpacked, relevant interventions need to be identified and agreed upon and they need to be implemented professionally. There also needs to be a strong focus on trustworthiness and attention is required for the following: joint planning and decision-making; agreeing on the role and contribution of all stakeholders; setting realistic and attainable goals; sharing information openly; a commitment to transparent decision-making; keeping promises, and helping partners to demonstrate competency.

Where trust between partners has been historically low then significant time needs to be spent to make these collaborations effective. If this is not undertaken then there is a real risk these well-intentioned vehicles for change will deliver low impact and key stakeholders will lose interest or withdrawal from the development process.

6.6.5 Outcomes

The Aqua for Sudan was implemented over a five-year period. Communities in Hamesh Koreib and North Delta now have improved access to water year-round as a result of the groundwater recharge structures and improved water points. People see improvements in physical infrastructure as one of the best ways to improve water security and a fundamental aspect of integrated water resources management. Households and communities speak about increased water availability as a direct result of these interventions.

However, it was also evident that water management in rural Kassala is becoming increasingly difficult, due to growing demands and pressures; and the breakdown of trust between communities – both within the catchment area and with government counterparts. Issues of trust include an adherence to agreed water management principles between upstream and downstream communities, which depends on a level of confidence that people will adhere to local rules when they are not being observed. Other reasons why communities struggle to manage their water resources included the absence of external support from state institutions over many years.

During the programme intervention it was discussed that communities had previously practiced traditional water management, although people's knowledge of the science of groundwater recharge and recovery was basic. Members of the recently formed catchment management committees in both Hamesh Koreib and

North Delta now have better knowledge and appreciation of water security. They now speak about the requirement for better hydrometric monitoring, sound water allocation, appropriate water laws and provision of physical infrastructure. They also recognise that water security cannot be achieved unless trust is developed between communities and government institutions. However, these partnerships require significant capacity building if communities are to receive new infrastructure and adequate external support when problems exceed their capabilities. These are problems that cannot be easily resolved, because many of these issues are entrenched. This means integrated water resources programmes in challenging environments need to focus on interventions that will achieve maximum impact. This challenges the view that the full scope of integrated water resources management can be achieved and local level interventions should focus on the most relevant and effective interventions. In simple terms practitioners must 'ensure the most appropriate or relevant intervention, based on local context' and implement effectively as a result of high-quality design, construction and supervision.

6.7 KEY LESSONS

6.7.1 Integrated water resources management needs to be unpacked when working in challenging environments

Integrated water resources management has been the dominant paradigm for many decades. However, it has been questioned for being overly complex and not necessarily focussing on solving real water management problems that people experience (Giordano & Shah, 2014). In order to make a difference it is necessary to unpack the approach in detail so it matches the context. There are dangers in focussing solely on the integrated water resources management principles and less on the actual water security problems that people face. The enabling conditions for sound water resources management are absent in Sudan and this means intervention approaches need to be simplified. The reality is INGOs need to pursue interventions that will achieve greatest impact and communities will likely undertake day-to-day work with minimal external support, until wider government support systems are strengthened.

6.7.2 Interventions must solve real water management problems that people experience

The challenges highlighted above means that implementing agencies must focus on solving water management problems that directly affect people's livelihoods. The only real way to build resilience and stimulate change is to ensure interventions are relevant to the local context and implemented to high professional standards to ensure effectiveness.

6.7.3 Developing a conceptual framework is an integral part of the integrated water resources management process

There are dangers in how integrated water resources management is implemented if practitioners are unable to visualise the key components of the approach being implemented. It can be a new approach for many organisations and practitioners, and 'full' integrated water resources management will be too ambitious for many fragile state contexts. Interventions may also overly obsess with the four guiding principles of integrated water resources management rather than focussing primarily on the local water security problems that people experience.

These risks can be avoided by developing a conceptual framework that identifies the key factors that need to be considered during implementation. Six key guiding principles identified as part of this work include:

(1) Address real water security problems that people experience – in order to ensure real demand. A first priority in Kassala is to increase water availability and access in the dry season months.

(2) Programme design and intervention should consider the key components identified by Perry (2008) and tailor them to the local context. Taken together they provide a systematic approach for implementation.

(3) Recognise what actions community-based institutions can realistically perform and be clear what exceeds their capabilities (see Table 6.1).

(4) Build trust and trustworthiness between community-based institutions and government agencies.

(5) Linked to Point 3, try to ensure appropriate and effective external support (technical, financial and management) to communities and catchment management committees.

(6) Focus on interventions that will achieve maximum impact. Given the inherent problems with responding to water service breakdowns and disruptions, high-quality infrastructure design and implementation is fundamental.

6.7.4 Community participation is essential but demands continuous external support

Community-based maintenance forms a central component of national water policy in Sudan. However, over the past 10–15 years the approach, which is widely applied in WASH service delivery, has been criticised globally for its ineffectiveness and the breakdown of rural water supplies. Despite the well-documented challenges community-based management remains relevant when working in particularly difficult and challenging contexts. In Kassala community representation and participation was fundamental in establishing catchment management committees from the outset. This is often because there is no other credible alternative. The critical issues are twofold: first, to ensure roles and responsibilities are clearly defined and appropriate to the internal capacities of community-based groups. Second, to try to ensure continuous external support when major problems arise. The following 12 categories represent areas of support required:

- periodic and refresher training on integrated water resources management by relevant state institutions;
- building trust between communities and government counterparts;
- mapping and analysing water security problems within the catchment area;
- hydrometric monitoring – collating, analysing and processing monitoring data to understand water availability and effectiveness of interventions;
- engagement with upstream and downstream water users to discuss water allocation and resolving any water disputes/conflict;
- defining local and catchment level water laws and rules for ongoing usage and prioritisation;
- determining roles and responsibilities across a range of stakeholders;
- technical assistance for planning water supply infrastructure (such as planning, design and high-quality construction;
- rapid technical assistance when physical infrastructure fails, including provision of spare parts;
- external support for when externalities arise (flooding or drought)
- continued financial support for water resource management activities and sustaining water supply infrastructure;
- support in documenting learning and sharing experiences.

6.7.5 Wider systems strengthening will take considerable time

To manage water resources effectively many institutions at multiple levels need to be involved. Community-based organisations can play an active role and should form part of wider catchment management committees. However, other 'professional' institutions at local, basin, national and transboundary levels also need to be involved and play their part responsibly. In fragile and conflict-affected settings, government institutions have often suffered years of under investment and technical support. This means they may possess knowledge and appreciation of water resources management, but they lack ability and practical experience. It also means any programmes that focus on systems strengthening are starting from a low base. There are numerous examples of institutions receiving well-intentioned 'capacity building' training, but this does not necessarily lead to improvements in the way institutions or agencies perform. In fragile state contexts training needs to focus on relevant interventions that will lead to some tangible difference in solving real water management problems that people experience. The only way to make a difference is to ensure the right training is provided professionally and technical support is continuous (multiple engagements), rather than short-term projects. Sudan has received decades of humanitarian and development support, and yet in many cases the development of national institutions has been limited.

6.7.6 Build trust and trustworthiness between stakeholders

Integrated water resources management at local, national and transboundary levels requires ongoing multi-stakeholder collaboration. Many organisations at multiple levels need to be involved. This can be a difficult process for many reasons: different water users at a local level may be competing for access to scarce water resources; there may be deep mistrust between communities and government institutions, water security problems are seldomly explicitly discussed; longer term planning is not put in place to build trust; organisational behaviour may not adequately change to build trust; and people may be unwilling to rely solely on another group's actions for their water security.

6.7.7 Communities and resilience

There are many rural communities in Sudan that receive limited support from federal or state levels of government. There are limited new rural water supply programmes, inadequate rehabilitation of water supply services and insufficient external support for CBM when infrastructure breaks down. At state level in Kassala, SWC are responsible for the provision of water supply services to urban, peri urban and rural communities. However, late and inadequate financial disbursements from federal government in Khartoum has led to SWCs growing inability to serve rural populations. Due to public pressure SWC mainly focus on areas with higher population densities (towns and cities); with INGOs and NGOs attempting to fill the service provision gap in rural localities. This could be considered a dereliction of duty by SWC and federal government.

Rural communities definitely need ongoing external support to operate and manage their water supply services. They also require continuous assistance to monitor and manage water resources. Despite these challenges, programme interventions must work with communities in order to improve water security. Key actions identified are summarised in Table 6.1.

6.8 CONCLUSION: BUILDING RESILIENCE AT COMMUNITY LEVEL

It would be naïve to think that short-term integrated water resources management interventions can solve all water security challenges in rural Sudan. There are multiple challenges in keeping water supply services working indefinitely and ensuring sound stewardship of water and land resources. However, this chapter highlights that community-based interventions are a necessity and should be persisted with. Here, we conclude with four key reasons.

First, rural populations, particularly those in remote hinterlands, remain particularly vulnerable to drought, water scarcity and increased poverty. There is a sheer absence of managerial, technical and financial support for rural populations and in such context's community-based interventions for both water resources management and water supply will remain the most feasible option for the foreseeable future.

Second, in the absence of external support, programme interventions must ensure they are relevant to the local context, and their implementation is effective. This can be summarised as *doing the right thing and doing it well*. These two key aspects will provide interventions with the best opportunity for success in the absence of external government support.

Third, there is ample evidence that community-based institutions can play an active role in monitoring and managing water resources. However, there are limits to the tasks that they are able and willing to perform. Wider systems strengthening across government institutions will take decades to achieve and there is a requirement to focus on the critical external support activities that will help to improve water security for rural populations.

Fourth, in complicated and complex settings there is a tendency to bypass government institutions. This is because there are risks associated with funding and government systems are not considered transparent. But at some point, humanitarian and development programmes must begin to engage with relevant government institutions so they can support community-based approaches. Many government institutions are in need of significant support but they are willing to play an important role. Despite a multi-year programme intervention the capacity of the Groundwater and Wadis Department in Kassala has changed very little. Much needs to be learnt about engaging with state and national level institutions and managing risk in post conflict settings if better resilience is to be achieved.

DISCLAIMER

The authors alone are responsible for the views expressed in this chapter and they do not necessarily represent the views, decisions or policies of Plan International or FCDO.

REFERENCES

Giordano M. and Shah T. (2014). From IWRM Back to Integrated Water Resources Management Available: https://www.tandfonline.com/doi/abs/10.1080/07900627. 2013.851521?journalCode=cijw20.

Perry C. (2008). The ABCDE of Water Management. Draft Guidance Note Contribution to Arab Water Academy Courses. The World Bank (2009) Water in the Arab world: Management perspectives and innovation. Available at: http://siteresources.world bank.org/INTMENA/Resources/Water_Arab_World_full.pdf.

Mokhtar K. and Day S. (2007). Darfur: Relief in a Vulnerable Environment. Available at https://learn.tearfund.org/-/media/learn/resources/policy/relief-in-a-vulnerable-envirionment-final. pdf (accessed 12th March 2021).

Chapter 7

Can and should refugees and communities that host them expect better performing and resilient water supply services?

Ryan Schweitzer[1], St John Day[2], David Githiri Njoroge[3] and Tim Forster[4]

[1]*WASH Engineer (independent), Milwaukee, USA*
[2]*Principal Consultant, IOD PARC, Edinburgh, UK*
[3]*WASH Officer, UNHCR, Addis Ababa, Ethiopia*
[4]*Relief and Development Specialist, Oxfam, Oxford, UK*

ABSTRACT

During the acute phase of an emergency the priority for humanitarian agencies is to rapidly establish water supply and other basic services (e.g. sanitation, hygiene, and solid waste) for people affected by disaster or crisis. However, the immediate response to an emergency is relatively short in duration, while the services, particularly water supply, often need to meet the needs of affected populations for many years. Often crises are protracted in nature and it is therefore important to understand how service performance evolves and whether service users are satisfied with the level of water supply. This is an important consideration because long-term sustainability may not represent an important part of initial thinking by humanitarian agencies. The United Nations High Commission for Refugees estimates the average time spent by a refugee in a camp is 10 years, while the average refugee camp remains for 26 years. Two questions arise: first, how will humanitarian agencies ensure emergency water supplies reach the desired performance levels; second, how will local institutions be able to manage, modify and finance the services that camp or settlement dwellers and host communities will depend upon. In this chapter the authors explore experiences from two country case studies and monitoring data extracted from ongoing humanitarian crises. The main conclusions are: service level enhancements are often slow to materialise and widespread efforts are required to raise performance levels.

Keywords: water supply, protracted emergencies, asset management, financial resilience

7.1 INTRODUCTION

Humanitarian crises can take many forms. They may be rapid or slow onset, and result from natural and technological disasters, armed conflicts or aggravating factors (such as climate change and drought). Those affected may be displaced within the country or forced to move across international borders. Host populations are also affected by displacement as greater demands are placed on their natural resources and services. Tearfund (2007) illustrates this risk in the following example:

> *The heavy environmental impact of prolonged displacement is degrading some of Darfur's most valuable agricultural land. Many IDP camps are built around agricultural market towns, which means that land degradation affects prime farmland, undermining livelihoods for both the displaced and the host population, affecting the crisis as well as the future recovery period.*

Each emergency also has a unique profile. Aside from root cause other important issues include: the underlying capacity of government and its national partners to respond, the baseline conditions prior to the emergency with regard to basic services, environmental conditions and the degrees of human and economic development and socio-political stability. These factors will affect the timing and quality of necessary transitions to more resilient and stable conditions. During the emergency to post emergency transitions there will be differing priorities and driving factors. Often, the transition process to stability can be drawn out and influenced by additional compounding crises (e.g. disease outbreaks, security issues, and funding gaps). These can serve to slow or even reverse progress.

During the acute emergency phase government and humanitarian agencies (United Nations, international or local non-governmental organisations (NGOs)) tend to focus on life-saving interventions and meeting critical needs of the affected through emergency funds and grants and temporary infrastructure, following humanitarian standards and rapid response mechanisms. As conditions stabilise, the scope and perhaps scale of programming may increase allowing actors to pursue other outcomes that include: mitigation against further risk, preparedness for future shocks, expanding health services, leveraging other sources of finance, and improving levels of service and user satisfaction. This transition is a dynamic and non-linear process; and in any given sector the rate of progress will be variable. Figure 7.1 presents a visual representation of dynamism inherent in this transition from acute emergency work to longer-term human development.

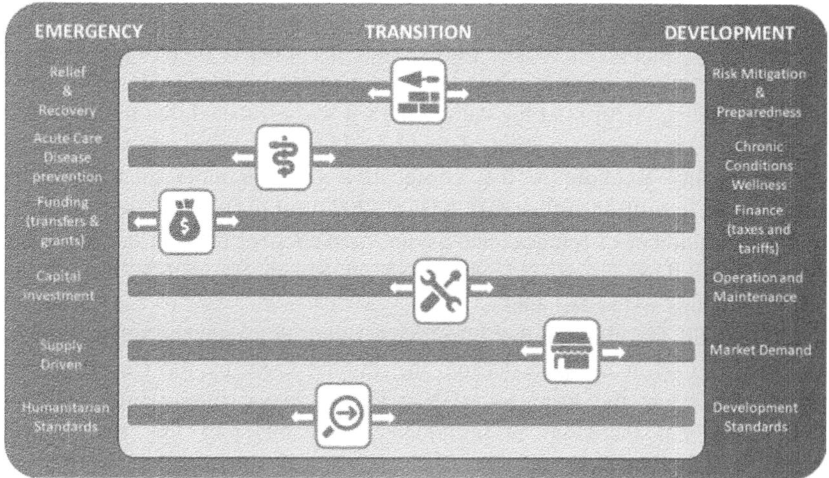

Figure 7.1 Representation of a number of factors or components which influence emergency and development programming.

Post emergency transitions may also be sudden or slow. Depending on the context people may return to their homelands, they may seek resettlement in another country, or apply for asylum in the host country. Many also migrate to other towns and cities within the host country or they may remain in refugee or internally displaced person (IDP) camps or settlements indefinitely, waiting for conditions to stabilise so that they may eventually return home. Arguably, as compared to an acute emergency response, the greater challenge may be to determine how to provide the desired levels of service to host communities and displaced populations that may remain for years, even decades, especially when host governments may be reluctant to build permanent facilities in host communities. Unfortunately, the scale of the challenge has increased dramatically in the past few decades.

Section 7.2 will describe the overall scale of the current humanitarian challenge. Then Section 7.3 will unpack the specific challenges for the transition from emergency to development for water supply services. Section 7.4 will present evidence of water supply services from two recent humanitarian emergencies which will be used to highlight five key areas for opportunity, that are further discussed in Section 7.5. Section 7.6 will provide a conclusion.

7.2 SCALE OF THE CHALLENGE

At the end of 2020 there were 82.4 million forcibly displaced persons globally (UNHCR, 2021c). Over the past decade, the trends in forced displacement have seen both an increase in the cumulative number of individuals forcibly displaced, as well as an increase in the duration of displacement and in the duration of the camps or settlements, which are established to support the basic needs of those displaced. For example, over the past decade the number of refugees living in camps has increased six-fold while the total number of displaced only doubled (UNHCR, 2020a, 2020b). At the end of 2018 it was estimated that the number of refugees in a protracted situation, defined as 25,000 or more refugees from the same nationality in exile for five consecutive years or more in a given host country, was 76% of all refugees (UNHCR, 2021c). Figure 7.2 shows the recent trends in the number of refugees and asylum seekers over the past decade, with a breakdown of those living in camps and settlements as well as those living outside of camps and settlements.

Although the absolute number of refugees living in camps and settlements has increased six-fold in the past decade, the amount spent by the United Nations High Commission for Refugees (UNHCR) on water supply and sanitation services has not increased proportionately and therefore this has required organisations to '*do more with less*' as the amount allocated by UNHCR per capita has considerably decreased. Figure 7.3 shows the amount UNHCR spent per capita on the provision of water supply and sanitation to refugees and asylum seekers in UNHCR planned and managed camps between 2010 and 2019. This does not account for any amount that is spent by operating or implementing partners, which may be considerable in some cases, particularly for the Syrian response in Jordan and Lebanon and the Rohyinga response in Bangladesh. Figure 7.3 also shows the population of refugees and asylum seekers in camps during the same period.

The majority of those living outside of camps or settlements reside in urban or peri-urban areas. Amongst these individuals, there are some that are targeted with assistance for meeting their basic needs such as food, housing and

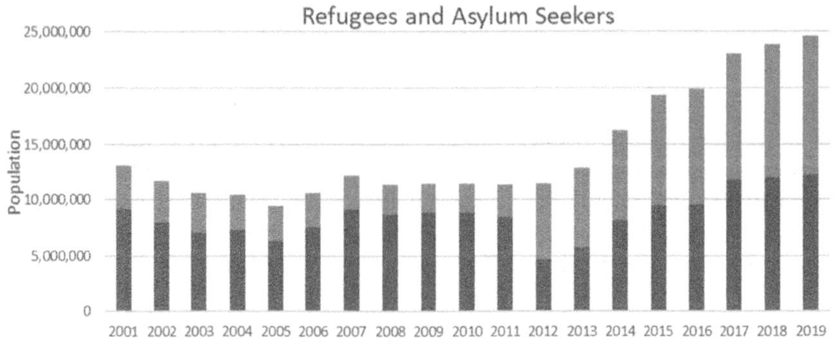

Figure 7.2 Number of refugees and asylum seekers over the past two decades, disaggregated by those living inside camps or settlements and those outside of these locations.

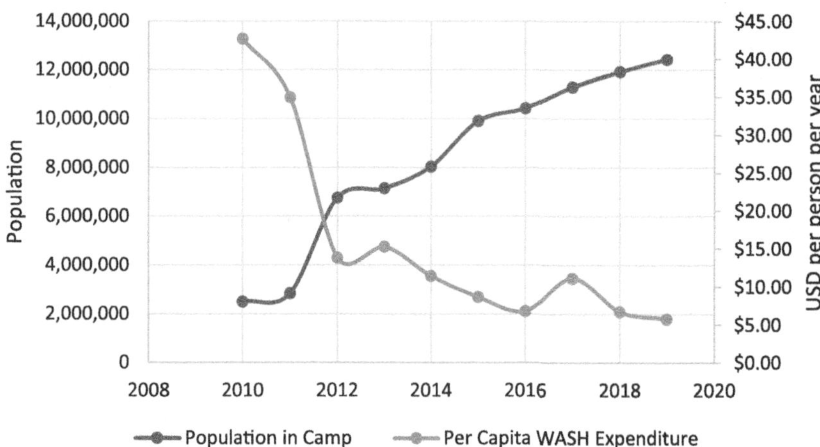

Figure 7.3 Population of refugees and asylum seekers living in camps and settlements and the per capita expenditure on water supply and sanitation made by UNHCR.

non-food-related items. This assistance is usually provided in the form of cash-based interventions. Indeed, UNHCR's policy is to pursue alternatives to camps, whenever possible, while ensuring that refugees are protected and assisted effectively. Despite this policy, in the past 10 years it has been necessary to establish over 180 camps and settlements to cope with the rapidly increasing number of refugees and asylum seekers (UNHCR, 2020a, 2020b). The vast majority of refugee camps and settlements are located in rural areas that are devoid of existing services and physical infrastructure. This makes service delivery extremely challenging for the humanitarian agencies.

As compared to urban areas, rural areas tend to have lower levels of access to safely managed water supply and also have a larger relative population accessing surface water sources. According to the Joint Monitoring Programme (JMP) of UNICEF and WHO in 2017, eight out of every ten people who lacked access to basic water supply services lived in rural areas, and nearly half lived in least developed countries (JMP/WHO, 2018). In addition, the JMP found that only 46% of rural inhabitants accessed improved sources which were free from faecal contamination versus 72% of urban inhabitants. There are numerous factors that attribute to the lower service levels in rural areas including:

- Limited capacity of local governments to provide services or monitor and regulate services provided by others (e.g. private sector).
- In rural areas, economies and markets are generally weaker than in urban areas, which has a number of implications including weaker supply chains, limited economies of scale, lower access to financial services, and less access to international markets.

- Job opportunities can be limited in rural areas with wages and salaries less competitive than in urban areas. This can lead to brain drain as people leave rural areas to seek opportunities in towns and cities.

As a result of these factors, it is important that humanitarian actors that are working to address the needs of the forcibly displaced also consider the impact on the host community and work to alleviate any additional pressure created by the displacement. Indeed, this is a key objective of the Global Compact for Refugees (GCR) (UN, 2018). The GCR has additional objectives of enhancing refugee self-reliance through increased livelihood and education opportunities and ensuring the freedom of movement and right to work for refugees. Achieving these objectives is critical for ensuring that water supply and other basic services can be transitioned from the typical humanitarian service delivery model to a more sustainable model which is in alignment with existing models operating in that context. Section 7.3 looks more closely at the issues in transitioning from emergency water supply to resilient solutions

7.3 TRANSITIONING FROM EMERGENCY TO RESILIENT WATER SUPPLY

7.3.1 What is the status quo

Following the onset of an emergency, if a camp or settlement must be established, often the land that is allocated by the host government is undeveloped and devoid of any civil infrastructure or access to public services. Therefore, in order to meet the immediate needs of the population during this acute emergency phase, the humanitarian agencies must provide life-saving assistance through temporary or transient infrastructure. This often includes water supply via water trucks or emergency sanitation facilities.

During this time period humanitarian agencies often work towards professional humanitarian standards (e.g. SPHERE) which target the minimum levels of services necessary for survival and to prevent disease. Depending on the scale of the displacement this work is usually led by the host government with operational activities often carried out by international NGOs (large displacements) or national NGOs (smaller displacements) or through direct implementation by the UN agencies who contract private companies. Figure 7.4 is an example of a common refugee response scenario in terms of expenditures on water, sanitation and hygiene (WASH). Dollo Ado camps in eastern Ethiopia received an influx of refugees from Somalia beginning in 2012. There was considerable initial investment in water trucking carried out by international NGOs followed by capital investment in water systems in 2013 and 2014, which is shown as a solid colour labelled 'CapEx' in Figure 7.4. This investment was mainly made by international NGOs with some direct implementation by UNHCR and local NGOs. What is also notable is the shift from capital investment to operation and

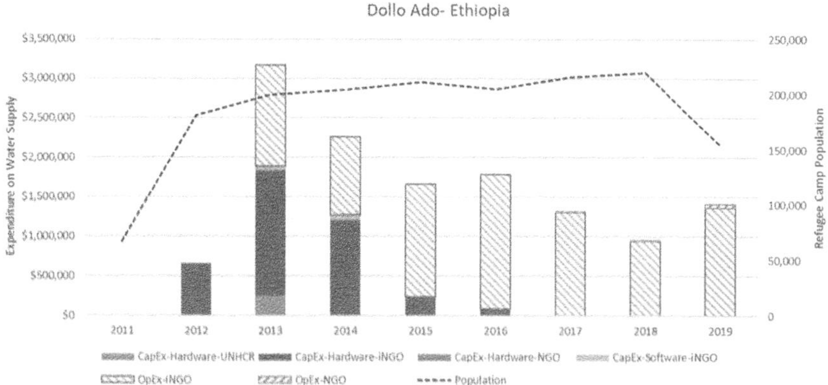

Figure 7.4 A rapid increase in the population of Dollo Ado required considerable capital investment to expand WASH services which was phased out over 5 years.

maintenance expenditure 'OpEx' (shown by the hashed boxes) which dominates as time passes from the emergency phase in 2012.

As the situation stabilises and the acute emergency phase concludes it is desired to transition to resilient solutions that would provide higher levels of service to the beneficiaries (displaced and host community). It is also important to address not only technical aspects, but also social, institutional, environmental, and financial aspects which contribute to the overall resilience of the water services.

It is critically important to begin planning for this transition as soon as possible. There are several reasons for this which include declining resources (financial and human resources) as well as declining social and political capital. In general, during the acute phases of an emergency there are considerable financial resources available and numerous actors available to support the response (international and national NGOs). International humanitarian organisations working through the Inter-Agency Standing Committee commitment can access various global funds such as the Central Emergency Response Fund (CERF) and Country-based Pooled Funds (CBPF). However, in general the funding available reduces as time passes. Figures 7.5(a)–(d) are illustrative of these trends in decreasing resource allocation over time. These figures show the population of refugees and asylum seekers living in camps and settlements across four countries (i.e. Kenya, Uganda, Ethiopia, and Iraq) over the past two decades, along with the expenditure on WASH by UNHCR (this excludes any overhead costs for UNHCR (e.g. staff, and staff expenses) and does not account for any additional expenditures made by other humanitarian organisations). In general, the trends show that as the forcibly displaced population grows (blue dashed line) there is a corresponding increase in WASH expenditure (solid green bars), but this expenditure decreases as time passes and media attention and donor earmarking

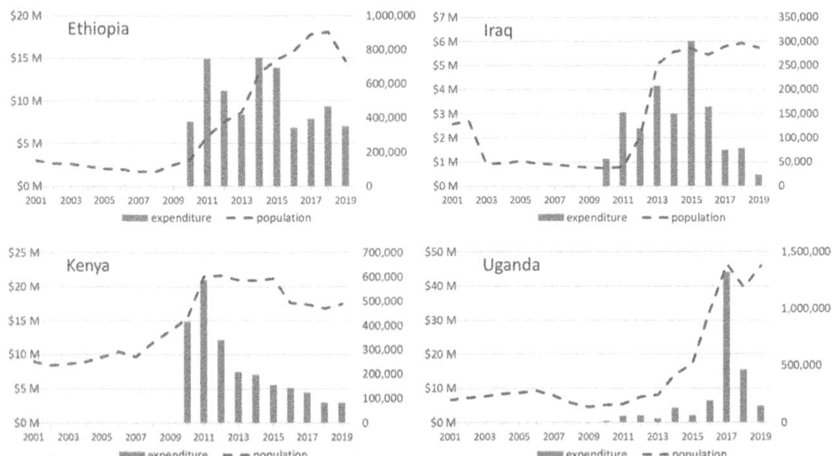

Figure 7.5 Refugee population in camps and settlements and expenditures on water supply and sanitation made by UNHCR in Ethiopia (a), Iraq (b), Kenya (c), and Uganda (d). Population data available from 2001 through 2019, while expenditure data are only available from 2010 until 2019.

wanes. However, if the displaced population does not decrease concurrently, it will mean less funding available per capita. There is evidence demonstrating that decreasing funding (per capita) can impact the ability to maintain or improve the level of water supply or other basic services, particularly those with high operation and maintenance costs (Moriarty *et al.*, 2011).

7.3.2 What is best practice

The year 2020 marked the milestone of one decade remaining until the deadline of the Sustainable Development Goals (SDGs), and at the current rate most fragile states will not achieve the established targets. Considerable attention in the literature has been given to WASH in fragile and conflict affected states, most notably with UNICEF's Water Under Fire series. However, globally there is very limited experience in successfully transitioning water supply services from emergency/humanitarian to sustainable and resilient service delivery. There is no roadmap or framework outlining best practice.

Section 7.4 looks at two specific examples of the transition from emergency to post emergency taking a deeper dive into the evolving service levels in this transition. The examples both result from the South Sudan civil war: the influx of refugees in Gambella, Ethiopia and the influx into Northern Uganda. These examples were selected for the following reasons:

- **Scale**: They represent large populations of forcibly displaced persons (in both cases the displaced populations were more than 200,000 placing them within the top five displacement situations in the past decade).

- **Availability of data**: Both situations were refugee emergencies and had data readily and publicly available. Specifically they had data on service level (proxies) throughout the entire transition.
- Both **host country governments are supportive** of the Global Compact for Refugees and the Comprehensive Refugee Response Framework and are facilitating the transition work.
- Development donors and **funding for the WASH response,** and specifically for the transition from humanitarian to development approaches, are available, including the establishment of public utilities to operate water supply systems.

7.4 LOOKING AT SOME EVIDENCE
7.4.1 Gambella, Ethiopia

The South Sudanese civil war, which started in mid-December 2013, led to a considerable influx of refugees into the Gambella Region (kilil) of Ethiopia. As of December 2019 refugees make up approximately 50% of the region's population (UNCHR, 2020a). Since the 2013 influx, a number of additional refugee settlements were established and other, existing camps and settlements were closed. Currently, there are three camps hosting refugees in the Gambella area, namely: Tierkidi, Kule and Nguenyyiel, which are located approximately 50 km east of the border with South Sudan. These camps combined host approximately 200,000 refugees. Kule and Tierkidi were established in March of 2014 and water service for both camps was provided via water trucking for approximately 24 months. Nguenyyiel Camp was established in October 2016 and water was also supplied via water trucking for 22 months (10/2016–8/2018) (UNHCR, 2019).

In 2014, UNHCR and UNICEF carried out a financial analysis comparing the ongoing water trucking costs for Kule and Tierkidi refugee camps and the options for more resilient water supplies. The determination was that trucking water for two years was equivalent to the capital investment for an entire water distribution network in the camps. As a result, 1 M euros in funding from the German government was secured for Phase 1 (12/2014 to 12/2015) for the development of water supplies, conduction line and storage reservoirs for Kule, Tierkidi and the host community (Birhanu, 2020).

In 2016, with additional refugee arrivals and in alignment with the Government of Ethiopia's commitments to the Comprehensive Refugee Response Framework, Phase 2 was initiated. This phase ran from 12/2016 to 6/2019 and had a cost of 6.5 M euros for an expansion of the water supply system connecting Itang Town, Thurfam community as well as Nguenyyiel Camp and the establishment of a public utility-based service provider. This was unique as it shifted responsibility for day-to-day operations from an NGO (International Rescue Committee – IRC) to a utility. Under the utility operation model there was an emphasis on cost-recovery, localisation, community participation, capacity development and coordinated partnership (UNICEF, 2018). The Itang Water Utility was formed in 2017 and took over operation of the system in December 2018 and received direct operational support from IRC until April of 2019. Currently the project is in Phase 3, the final phase of the project which includes an estimated 10 M euros earmarked for an expansion of the system for 30,000 refugees and funds to optimise the system and improve the capacity of the Itang Utility (Birhanu, 2020). The current Itang Water Supply System provides water to almost 840,000 people (48% refugees and 52% host community) and the utility is collecting a tariff of 16.91 ETB (0.40 USD) per cubic metre of water delivered (Birhanu, 2020).

The project faced a number of considerable challenges including vandalism to equipment and infrastructure, reduced operating times due to security risks, difficulty accessing spare parts, logistics challenges, human resource and administrative challenges (e.g. non-competitive salaries for utility staff, ineffective board) (Birhanu, 2020). Considering these challenges and the scale and complexity of the operating environment (e.g. remote area, relatively low service levels prior to influx, high cost of service delivery), the achievements accomplished to date may be commended. However, the transition from humanitarian to development programming has, at times, come at a high price borne by the refugees and host community in the form of services that do not meet even the lowest humanitarian standards, much less the national targets recognised by the government (25 lpcd (litres per capita per day) for refugees and 40 lpcd for the host community; GoE, 2016).

To get a longitudinal look at the water services provided in these camps since the influx, data were collated by the authors from publicly available data sources and reports including: UNHCR's WASH monitoring system (UNHCR, 2021a), situation reports from UNCHR, 2021b, IRC, OXFAM and other partners, and data compiled by the Itang technical working group. Figure 7.6(a)–(c) shows the trends of water supply (expressed as litres per person per day) together with the volume of water provided via water trucking. These graphs show that since the initial emergency onset in 2014 (or the establishment of the camp Nguenyyiel in 2017) the general trend in quantity of water provided per person per day has been flat, with no improvement in the quantity of water provided to the refugee camps. What is concerning is the substantial time periods when the supply drops well below minimum humanitarian standards of 15 lpcd. This is most evident during the time period since the utility has taken over operation of the system. (It is worth noting that following the upgrading of the water supply infrastructure (2016–2019) and establishment of a reticulated system, the quantity of water provided was tracked daily while at the outset of the emergency the quantity delivered was tracked weekly. Although the measurements for the reticulated system are more precise, these do not account for any loss within the distribution networks of the camps.)

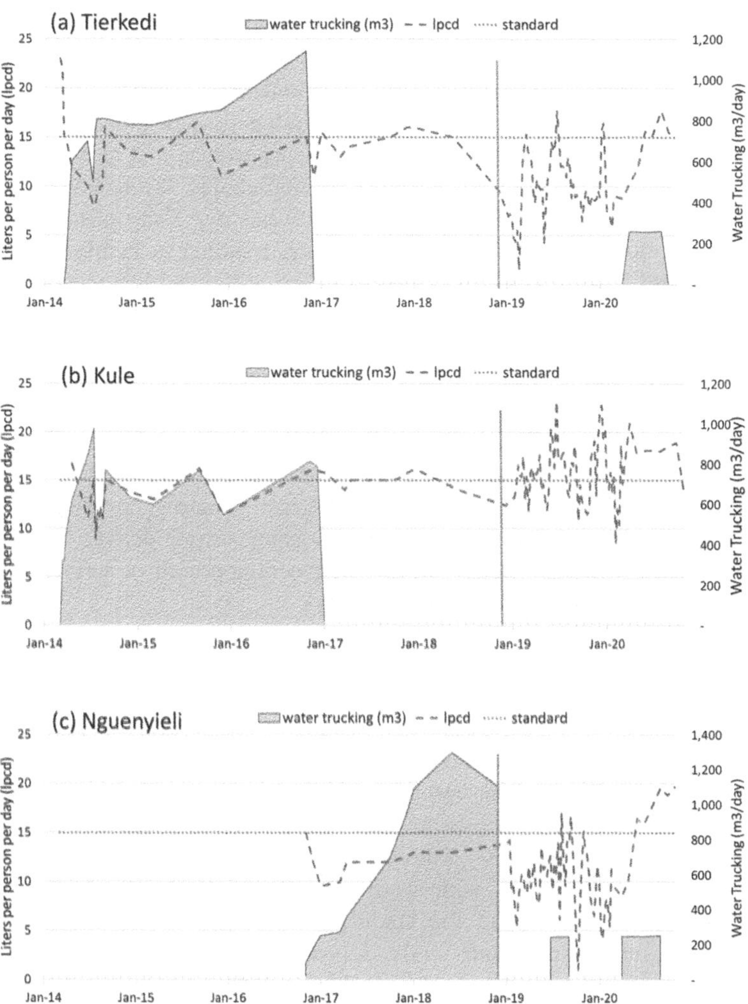

Figure 7.6 Water provided via water trucking (cubic metres per day) and the average quantity of water available to refugees and asylum seekers (litres per person per day) in Tierkidi (a), Kule (b), and Nguenyyiel (c) camps. The green line identifies when the Itang Water Utility took over operation of the systems.

These data presented in Figure 7.6(a)–(c) were corroborated by household surveys and other qualitative and health data collected by WASH partners. For example, a survey conducted by Oxfam in November 2018 immediately before the system was handed over to the utility revealed that just 20% of the 387 households surveyed had access to 15 litres per person per day (l/p/d) and two thirds of households interviewed perceived water supply to be inadequate. A total of 55% of households received <10 l/p/d and 23% were able to access less than 5 l/p/d. Service disruptions lasting up to 8 days were also commonplace, forcing people to access water from more distant and unprotected sources (OXFAM, 2019).

Focus group discussions (FGDs) carried out during the same time frame revealed considerable concerns about protection risks facing women and girls who had to travel long distances to collect water for their households. Women also reported difficulties in carrying a sufficient amount of water over long distances to the alternative source located 2–3 hours walk away, resulting in families having to limit their water consumption and habits – including limiting hygiene practices – when having to use these alternative sources. In addition, FGDs with children revealed that three quarters experienced diarrhoea after drinking water from the pond (OXFAM, 2019). Data collected by Medecins Sans Frontieres corroborated the increase in diarrhoea during the periods when water supplied by the system dropped below the desired targets.

These data suggest that even five years on from the emergency there were still considerable challenges to meeting humanitarian standards in the Gambella camps, despite the shift from the typical emergency service delivery model (i.e. NGO operation of infrastructure) to a development approach of service delivery through a utility (i.e. Itang Town Water Utility).

7.4.2 Northern Uganda

Another example of humanitarian to development transition in water service provision is the South Sudanese situation in Uganda. Since 2017, over one million refugees have sought refuge in Uganda, making it the third largest refugee-hosting country in the world after Turkey and Pakistan with over 1.4 M refugees as of March 2020. The Government of Uganda has been very progressive in its response with a strong commitment to promote refugee self-reliance and inclusion in the country's development planning. This has resulted in a fundamental shift in the approach to service delivery in the 12 settlements in the country by linking the traditional humanitarian response to long-term development approaches. For WASH practitioners this means transferring operation of WASH services from NGOs to private and parastatal Ugandan water utilities. Humanitarian actors have been successful in making a quick transition away from temporary services such as water trucking and community sanitation facilities. Monitoring data were obtained for twelve refugee settlements in Uganda including six settlements in the north hosting the refugees from the 2017 influx, namely: Adjumani, Bidibidi, Palabek, Palorinya, Rhino, and Impevi. Figure 7.7(a)–(f) shows this transition for the six settlements in the north which house nearly two thirds of refugees living in camps and settlements in Uganda.

Figure 7.7(a)–(f) shows that water trucking was scaled back, by at least 80% in the first 1.5 years, and in four sites had been completely eliminated by June 2019.

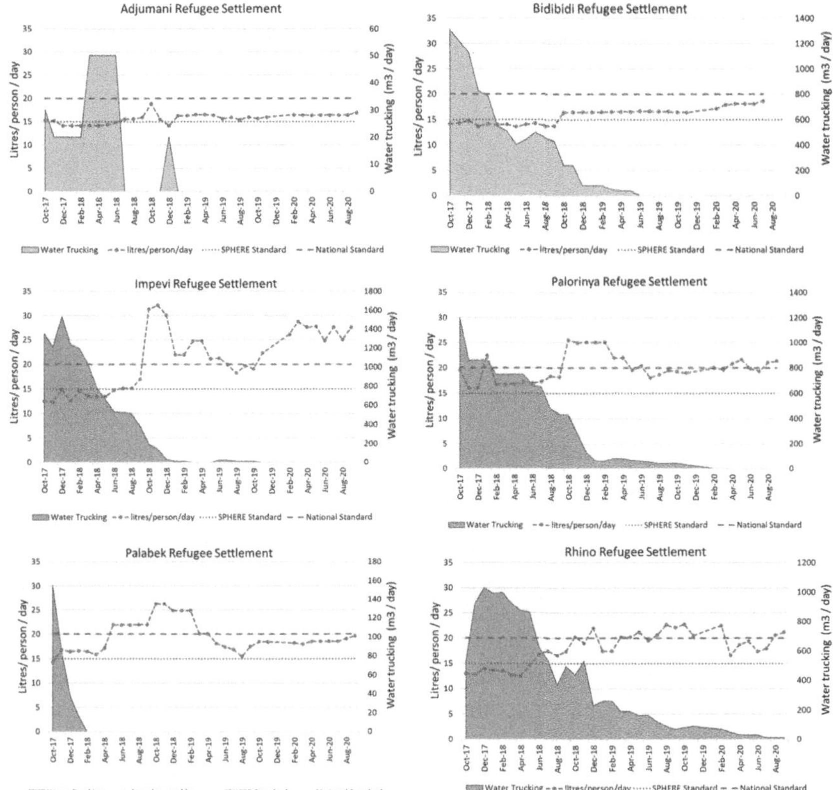

Figure 7.7 Quantity of water supplied per person per day and water provided via water truck (cubic metres per day) in six refugee settlements in northern Uganda: Adjumani, Bidibidi, Impevi, Palabek, Palorinya, and Rhino.

These achievements alone are very laudable. Analysing the progress in the quantity of water supplied over the 3-year period (October 2017–August 2020), using a linear regression shows that since the influx there has been no change (i.e. less than 1% change $+/-$) in five of the six camps. Only in Impevi has the trend shown an increase in the quantity of water supplied over the 3-year period for which there is data, going from 12.5 to 27.5 l/p/d with an annual average increase of 30%. Although there has not been an increase in services in most of the camps, there has been general success in meeting SPHERE emergency standards of 15 l/p/d. During the period of record 81% of the time the sites were at or above SPHERE standards.

In March 2020, the National Water and Sewerage Corporation (NWSC) took over management of water supply in Rwamwaja in Kamwenge District (see Figure 7.8). This represents the first transition to a local/national utility within a

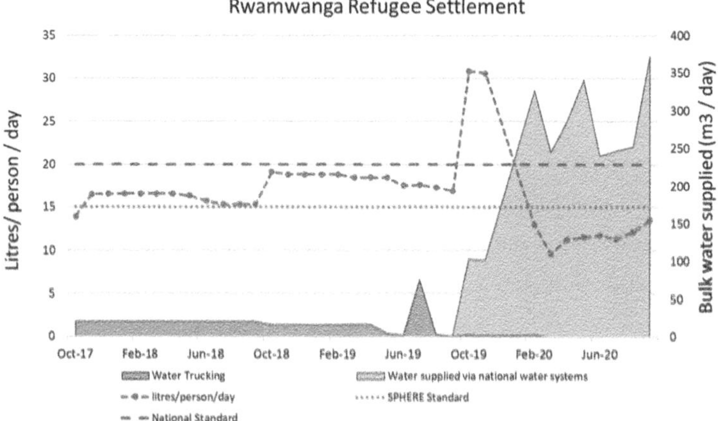

Figure 7.8 Quantity of water provided (liters per person per day) and bulk water (cubic metres per day) supplied via water trucking and the national water systems.

refugee settlement in Uganda. This settlement hosts 71,707 people, which is equivalent to 5% of the refugee population in the country (UNHCR, 2020a, 2020b). The process began in September 2017. Similar to the experiences seen in Ethiopia, the transfer to the utility has seen an initial drop in service levels from 30 l/p/d to less than 15 l/p/d. Some of this can be attributed to reduced focus on repair and maintenance of handpumps and point sources, which is not an area of strength for utilities.

Other systems in Uganda are slated to be handed over to NWSC or umbrella authorities. As of November 2019 the Northern Umbrella for Water and Sanitation (NUWS) took over management of water schemes in Nyumanzi settlement in Ajumani and has expanded to cover an additional seven water schemes including Rhino and Bidibidi settlements. These transitions are an important milestone for humanitarian organisations as they represent an opportunity for resilience with a permanent local/national authority operating the services. Transition of management of WASH services from NGO partners to Utilities in Uganda is anchored within an overall framework under the Ministry of Water and Environment (MWE). A Water and Environment Sector Refugee Response Plan (WESRRP) for refugees and hosting populations was developed, bringing together actors in development and humanitarian space. A Refugee Response Sub-Group was established and incorporated within the MWE coordination structure. A secretariat was created within MWE to oversee implementation of the WESRRP with a steering committee comprising of diverse stakeholders in development, humanitarian and private sector spaces. In addition, development donors are commissioning a consulting firm to carry out a detailed assessment of financial costs, operational performance, and user satisfaction before and after transitions.

7.4.3 Discussion

To date, emergency to post emergency transitions have not been well documented and as a result evidence-based best practice guidelines have not been established. Each emergency poses its own unique challenges, and many of these are not easily addressed. In emergency response, services are often provided free of charge and managed by external agencies (UN or INGOs). In the long term, national and local institutions, with vastly different capacities and resources, need to take on management responsibilities.

It is important to note that humanitarian agencies are working in some very extreme and challenging environments with crises occurring in localities where the optimal conditions or 'enabling environments' in terms of market access, governance, and social and political development do not exist. The authors note the transition efforts described above in Sections 7.4.1 and 7.4.2 are laudable. Furthermore it is encouraging to see the interest from development donors and host governments in Ethiopia and Uganda to support this transition with finance and political buy-in. However, it is clear that increased efforts are needed to improve the planning, implementation, monitoring, and support of these transitions. Furthermore, the authors believe that efforts should be made to decrease the timeframe for initiating the transition. For example, the case studies presented saw the transition begin four years (Ethiopia) and three years (Uganda) after the emergency. The important question is: can planning for this transition begin immediately during the emergency phase?

The evidence presented shows that there is limited improvement in service levels (the proxy of which presented herein is quantity of water per person per day) while services are operated by humanitarians. It is, perhaps logical, that humanitarian agencies would be working towards humanitarian standards (e.g. 15 l/p/d), and given the constraints (financial, administrative, technical, etc.) this may be expected and even acceptable. However, there is considerable evidence that increased access to improved service levels has benefits in human health, economic productivity, and user satisfaction. Therefore accelerating the transition should be the objective.

Importantly, the success of emergency to post emergency transitions cannot only be gauged in financial terms (e.g. money that is saved from water trucking, or the financial sustainability of the resulting system). It also needs to be measured in terms of the impact on the health, well-being, and satisfaction of the consumers (displaced persons and host community) that are accessing the services. Humanitarian agencies may also not be able to address the full scope of challenges present in such transitions, which means they need to identify interventions that may have the greatest positive impact given their constraints.

The following sections explore the five key opportunities, which have emerged from these case studies and the data presented, and also a series of desk reviews and investigations into the outcomes of emergency to development transition in six different countries: Bangladesh, Ethiopia, Ghana, Jordan, Nepal, and Uganda (Day *et al.*, 2020).

7.5 FIVE AREAS FOR IMPROVING POST EMERGENCY SERVICE DELIVERY

Once the acute phase of an emergency is declared over, systems for providing adequate water supplies will be needed for the foreseeable future. Logically this demands clarity and commitment concerning medium- and long-term water supply solutions. This requires support from the host government and other advocates who are willing to consider different water service options. An important consideration will be whether integrated services can be provided for both displaced and host populations and whether there are opportunities to strengthen the capabilities of national or local water utilities that may be present and willing to take over service delivery responsibilities. Other transitions will also be required. For example, humanitarian agencies will likely need to modify emergency water supply networks and they must ensure services are *relevant* to the local context and professionally implemented, in order to be *effective*. Reporting on service levels will also need to improve and emergency agencies will want to ensure adequate, safe and reliable water services, with minimal service disruptions.

For good reasons, those that manage humanitarian crises may also wish to rationalise the number of agencies involved in service delivery. Or they may wish to handover operations to public or private utilities. In order to do this the level of skills and resources available will need to be determined because they may be significantly less than those of humanitarian agencies. Even if the physical infrastructure is functioning effectively, there may not be sufficient data to determine the actual life-cycle costs to be able to sustain service levels. This is a common knowledge gap when water supply systems are handed over. Other issues to consider when services are handed over to other entities include: inaccurate or incomplete information concerning water supply networks; out-of-date asset information; absence of asset registers; availability of asset failure and repair data; and information on the cost of asset renewal. Day *et al.* 2020 outlines these challenges to service delivery in more detail. The following sections identify five areas that could help to remedy these challenges.

7.5.1 Independent assessments

The introduction of independent assessments could provide more convincing evidence for long-term planning. This is because emergency WASH services are often provided by humanitarian agencies (UN or INGOs). At least in the earliest days of an emergency, agencies conduct their own needs assessments and secure funding for implementation, either directly or by contracting other local agencies and institutions. This will often take the form of establishing emergency water supplies, and excreta and solid waste disposal, promoting safe hygiene behaviours, and so on. Agencies assess humanitarian needs, implement services directly and report on impacts. However, once media attention wanes and funding diminishes (as shown in Figure 7.3), the international agencies may move on to another global crisis or considerably downscale their operations. In undertaking independent assessments involving local institutions (e.g. municipalities, local government offices of water, health, education, and other line ministries, local regulators, etc.), who typically may be left responsible for modifying, managing and financing services long term, it is reasonable to assume that planning will factor in transitions at an early stage.

Another important consideration is whether current funding models perpetuate refugee or IDP camps. Barder (2018) highlights that refugee camps may be restrictive places with people unable to come and go as they wish and seek new employment opportunities. This is evidenced from the ongoing Rohingya crisis in Cox's Bazaar, Bangladesh. The provision of services to these camps is also dependent on aid agency funding models. Humanitarian agencies must recoup their own operational and logistical costs and there may be little incentive to move beyond direct implementation until funding streams diminish. This means humanitarian agencies do not currently rely on independent assessments or regulation to determine which long-term WASH systems are most viable and what service levels displaced people and host communities desire.

7.5.2 Asset management

With the projected reduction in official development assistance due to COVID-19 and the decreased funding to WASH by some agencies (e.g. UNHCR Figure 7.2), it will be crucial for humanitarian agencies involved in service delivery to adopt asset management planning practices into their operations. By engaging in asset management it will be possible for agencies to strategically manage their assets and better understand the minimal resources needed to maintain the desired service levels. This is vital information for service providers and investors that should accompany any handover of services.

In its simplest form asset management is a process that should enable service providers to meet the desired level of service in the most cost-effective manner. It is a systematic process to determine (a) the condition and risk associated with the assets, (b) which assets are a priority for renewal in the short to medium term and (c) the realistic level of funding required to renew assets on a continuous basis. The word 'continuous' is important because the process of asset management planning, and the related finance, is not a one-off activity but rather an ongoing process.

There are five basic elements to asset management (Boulenouar & Schweitzer, 2015). The first is that humanitarian agencies should maintain an asset inventory

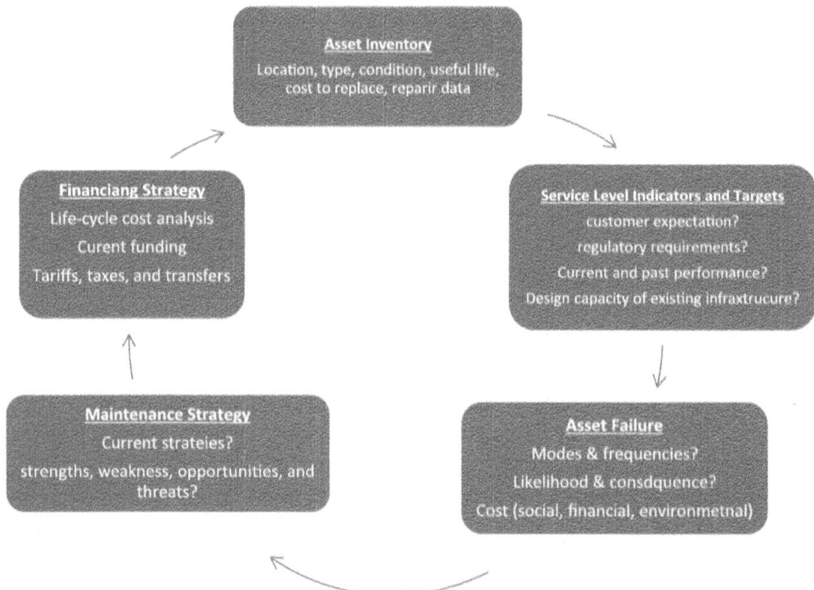

Figure 7.9 Five elements to asset management. Adapted from Boulenouar and Schweitzer (2015).

or register. This should include geo-location details of all infrastructure combined with the current condition, design life, and financial value or replacement cost. The second element is to determine the service level standards and targets. The third is to identify and analyse the modality of failure of each asset as well as the likelihood and consequences of failure including costs (such as financial, social, and environmental). The fourth is to establish the maintenance strategies for each asset or system of assets. These need to take into account personnel and capital budget accounts. The fifth and final requirement concerns the collection and analysis of data for identifying the life-cycle costs to maintain the service level standards and devising a financing strategy to maintain (and replace when necessary) assets at the desired level. Figure 7.9 gives an overview of the types of questions that are answered as part of each stage or element of the asset management plan, which is a permanently occurring and evolving process.

7.5.3 Service level targets

Logically and by reference to well established service delivery practices, once SPHERE minimum humanitarian standards have been achieved, long-term service standards and targets need to be determined and agreed upon. These may include indicators around the quality, quantity, reliability, hours of service, and access locations to be provided to service users. If this significantly exceeds minimum humanitarian standards, then new or improved assets may be required. Data in Figure 7.7(a)–(f) show that often service levels, in this case the quantity of water provided per capita, are erratic or limited improvements are achieved with the passage of time. However, if the desired level of service is not determined it will be impossible to decide the most cost-effective method for service delivery. Thus determining the agreed service level is fundamental for future investment planning and engagement with donors and financing agencies. Where funding declines or remains limited, service providers must prioritise asset renewal. There is also a requirement to consider both the condition of assets and their importance to the overall performance of the water supply network. For example, a pump may be considered to be in a poor physical condition, but still performs at the required head and flow. Thus it could be said to still perform well. However, a water supply pipe may be considered to be in a good condition but silt deposits, scaling, or biofouling result in poor performance. Thus, the operator needs to consider (a) the probability of asset failure and (b) the consequences of asset failure, both of which are informed by the asset register and data base.

7.5.4 Costing and financing of services

Finance is critical if post emergency water supply services are to remain resilient. Prior to handover, humanitarian agencies should estimate the true cost of service delivery, which includes minor operation and maintenance costs as well as asset replacement costs (sometimes referred to as capital maintenance costs). This is necessary both operationally, so assets are understood and can be managed, and strategically – to inform long-term financial planning and investment.

In addition, as mentioned previously, the transition from emergency to long-term development programming may involve well-resourced humanitarian agencies handing over to relatively weak local institutions or utilities (i.e. with limited resources, equipment and personnel). In these situations a considerable transition phase is required during which local institutions are supported to achieve the desired service levels (users) and operational performance (service providers). The timeframe of this transition is likely to be inversely related to the strength of the local institution (i.e. weaker institutions will require longer transition periods with greater support). In some cases it may be necessary for the humanitarian sector (and its funding or financing agencies) to expect a protracted transition period over many years and perhaps as long as a decade. For humanitarian agencies, of which many rely on annual budget and planning cycles, this may be a difficult task.

7.5.5 Capacity

Any emergency to post emergency transition will involve detailed discussions regarding who, realistically, can take on the long-term responsibility for managing and operating water supply services. Options may include: investing in existing water utilities; establishing new entities from scratch; handing over services to government agencies or continuing with the operation by humanitarian agencies (often with the objective to rationalise the number of agencies and prioritise strengthening of local agencies). Another trend that has been observed in East Africa is to establish rural water utilities. However, none of these options are straightforward and it is critical to assess what is the most viable long-term solution. In determining which service delivery approach to pursue it is important to consider the various factors that will impact the resilience of the services. In assessing receptive capacity it is reasonable to assume that humanitarian coordinators will document or benchmark the key attributes that need to be in place. This would serve as a reference for assessing receptive capacity and determining what long-term training and support is required. A conceptual framework for qualitatively assessing the service delivery options is presented by Day *et al.* (2020), however, in addition it would also be worthwhile developing quantitative metrics for assessing and tracking utility performance over time.

AquaRating is one such international rating system for water and sanitation utilities which focuses on the challenges water and sanitation utilities face in a comprehensive manner. It evaluates their performance through indicators and management practices, establishing an international standard. Importantly it relies on information verified by independent auditors accredited by AquaRating.

The primary purpose of using AquaRating in post emergency situations is fourfold:

- To establish an objective baseline of the performance of emergency water supplies provided by multiple agencies.
- To assess the capability of the local water company or institutions that will take on responsibility for managing.
- To inform improvement action plans during the transition period.
- To enable improvements to be measured over time.

The rating system assesses a range of competencies such as: service quality, investment planning and implementation efficiency, operating efficiency, business management efficiency, financial sustainability, access to services, corporate governance and environmental sustainability. The system is considered relevant because it recognises that improvements will inevitably be necessary and it focuses on increased accountability to service users – both consumers and customers. Thus it pinpoints critical areas for improvement and change management as well as opportunities.

AquaRating is one assessment system, but other approaches exist (such as customer survey benchmarking, establishing core monitoring indicators or process benchmarking). What is important is that there is stakeholder engagement in the process and transparency with regard to what the indicators are, how data are collected, collated, and reported, and what actions are taken to address any gaps or deficiencies.

7.6 CONCLUSIONS

While there is a strong desire to integrate refugee and host community water supply services and handover to a professional water utility this may not be a straightforward transition. In many contexts this may not be a viable option, and it may take years to build the enabling environment conditions for another entity to operate and manage water services. In many post emergency situations, in order to maintain and improve service levels, it is reasonable to think that humanitarian agencies (UN or INGOs) will need to have a long-term presence so that service levels can be incrementally improved.

For this to be achieved, humanitarian agencies also need to improve their own performance. To ensure emergency to post emergency transitions are viable then standard measures like independent assessments, asset management planning and agreed service levels are particularly useful. Furthermore, knowledge on the true cost of service delivery and asset replacement are crucial for future investment planning and engagement with international donors.

The most important conclusion from this paper is that better data are needed from post emergency situations. Standardised data for assessing service performance levels is needed to inform practitioners, decision-makers, donors/financing agencies and service users. If we want to make post emergency water supplies more resilient, we need to define service levels and establish detailed asset management plans. Data presented in this paper show that these are pressing issues to address. It is also necessary to determine the true cost of service delivery and assess the performance of any other entity that may assume greater responsibility for service delivery. Rapid decline in service levels or simply handing over services to another entity with minimal ongoing support and accountability to service users is unacceptable.

In Ethiopia and Uganda the humanitarian actors (UN and civil society) are working closely with the national line ministries and development donors to improve the outcomes of the transition. The experiences and lessons learned from these transitions will be critical to building a body of evidence for future transitions. It is evident that forced displacement will continue as will lengthy durations of displacement and therefore there is a need to have in place plans and procedures for transitioning to resilient service delivery approaches.

REFERENCES

Birhanu Y. (2020). Integrated water supply model serving refugee communities and host communities in Gambella. Presentation at the symposium on 'Climate resilience systems approaches for small town WASH services in Ethiopia'. 3/12/2020.

Boulenouar J. and Schweitzer R. (2015). Infrastructure Asset Management for Rural Water Supply Briefing notes series – Building blocks for sustainability. Available at https://nl.ircwash.org/node/78983.

Day S. J., Forster T. and Schweitzer R. (2020) Water Supply in Protracted Humanitarian Crises: Reflections on the Sustainability of Service Delivery Models. Available at https://oxfamilibrary.openrepository.com/handle/10546/621043 (accessed 21 September 2021).

Government of Ethiopa (2016). Growth and Transformation Plan 2.

Moriarty P., Batchelor C., Fonseca C., Klutse A., Naafs A., Nyarko L., Pezon C., Potter A., Reddy R. and Snehalatha M. (2011). Ladders for Assessing and Costing Water Service Delivery. IRC International Water and Sanitation Centre. Available at https://www.ircwash.org/sites/default/files/Moriarty-2011-Ladders.pdf (accessed 21 September 2021).

Tearfund (2007). Darfur: Relief in a Vulnerable Environment. Available at https://learn.tearfund.org/-/media/learn/resources/policy/relief-in-a-vulnerable-envirionment-final.pdf (accessed 21 September 2021).

UNHCR (2019). Global Trends Report – Forced Displacement in 2018. Available at https://www.unhcr.org/statistics/unhcrstats/5d08d7ee7/unhcr-global-trends-2018.html (accessed 21 September 2021).

UNHCR (2020a). Global Trends Report – Forced Displacement in 2019. Available at https://www.unhcr.org/statistics/unhcrstats/5ee200e37/unhcr-global-trends-2019.html (accessed 21 September 2021).

UNHCR (2020b). UNHCR Presence and People of Concern. Available at https://unhcr.maps.arcgis.com/apps/webappviewer/index.html?id=2028db44801d43fe8eb49321eea19285 (accessed 21 September 2021).

UNHCR (2021a). WASH Monitoring System. Available at https://wash.unhcr.org/wash-dashboard-for-refugee-settings/ (accessed 21 September 2021).

UNHCR (2021b). UNHCR Operational Portal. Available at https://data2.unhcr.org/en/country/eth (accessed 21 September 2021).

UNHCR (2021c). Global Trends Report – Forced Displacement in 2020. Available at https://www.unhcr.org/flagship-reports/globaltrends/ (accessed 21 September 2021).

UNICEF WASH FIELD NOTE / FN/07/2018: Improving WASH Service Delivery in Refugee Settings: The Experience of Localization and Professionalization of WASH Services in Ethiopia.

United Nations (2018). Global Compact for Refugees. Available at https://www.unhcr.org/5c658aed4 (accessed 21 September 2021).

Chapter 8

Solar-powered water systems for vulnerable rural communities: Alleviating water scarcity in Iraq

Mohammed Al-khateeb[1] and Ali Alkhateeb[2]

[1]*Researcher and Freelancer, Baghdad, Iraq*
[2]*Chief of WASH, UNICEF Iraq, Baghdad, Iraq*

ABSTRACT

Deteriorating water quality and decreasing water quantity are causing a water crisis in Iraq. The crisis is having a profoundly negative impact on people's livelihoods and on the economy. In the most water-stressed areas, vulnerable people have had to move from rural areas where water is scarce to urban areas, placing additional pressure on the water supply.

To mitigate the impact of water scarcity on the most vulnerable people in rural areas, the United Nations Children's Fund has worked in partnership with the Iraqi Water Authorities on a programme to increase access to more resilient water services in some highly vulnerable rural and conflict affected areas of Iraq where water services are unreliable. One major contributing factor to the problem of access to water was the unreliability of the electrical supply, particularly in the summer months. The programme identified that an alternative to grid electrical power was needed to achieve a more reliable source of energy for water provision.

The programme installed solar-powered water systems in two vulnerable districts in northern Iraq: Shekhan district, Ninewa, and Makhmur, Erbil. These systems are now providing sustainable, predictable and reliable water services to two vulnerable districts which had previously suffered extensively from power shortages and service interruptions. The water from the new solar-powered systems provides access to safe water for refugees and internally displaced people, as well as local communities. Importantly, the programme has increased water conservation and efficiency and helped to strengthen community resilience. It has also highlighted the need for adaptive and innovative technological solutions, which can support more effective disaster response and recovery.

Keywords: water scarcity, solar energy, rural water supply.

8.1 INTRODUCTION

The limited availability of water in sufficient quantity and quality is a daily challenge facing millions of people in Iraq, where water scarcity is both a physical and economic issue. Water is scarce due to an arid climate, but also because there is a lack of capacity and investment to improve water services and, as a result, water demand is not being met. Iraq has been identified as being under high water stress and is facing an acute water crisis (Maddocks & Luo, 2015). Water stress has manifested itself in a number of ways in the country, with water shortages and water pollution leading directly to violent protests, such as those experienced in Basra in 2018, and displacement of people as a direct result of water scarcity (UNICEF, 2019).

The water crisis in Iraq threatens its economic development and the achievement of its commitments to achieve the UN Sustainable Development Goals (SDGs). According to the last WHO-UNICEF Joint Monitoring Programme (JMP) report in 2019, while 91% of the rural population have access to at least a basic water service, only 46.5% have access to a safely managed service. It is a similar situation for sanitation access; the JMP shows that 88% of the population have access to at least a basic sanitation service, but only 45% have access to a safely managed sanitation service (WHO & UNICEF, 2019a, 2019b).

Poor access to safe water and sanitation services poses significant health risks and increases the risk of waterborne diseases, especially among vulnerable groups such as children and women (UNICEF in Iraq, 2019). Access to water and sanitation services varies significantly across different populations in the country, with some communities affected disproportionately by poorer access to services. According to the Government of Iraq, poverty is geographically concentrated in certain areas where it is deeply rooted and chronic and is associated with living in rural areas, amongst other factors (Ministry of Planning, 2010a, 2010b). According to the JMP data for safely managed water services, 64% of the urban population have access to a safely managed water service, compared to only 46.5% in rural areas (WHO & UNICEF, 2019a, 2019b).

Climate change and water scarcity pose a severe threat in Iraq, with large parts of the country facing potential desertification. As a direct result of water reduction into parts of Iraq, it has been estimated that 250 km^2 of Iraq's land is already becoming barren each year, causing a significant negative impact on livelihoods in rural areas (Ministry of Planning, 2010a, 2010b).

Water scarcity is already severely and directly affecting communities: in 2019, over 5000 families were displaced from the southern region of Iraq, particularly from Basra, Missan and Thi-Qar governorates, due to a lack of water (United Nations, 2018). The International Organization for Migration (IOM) has similarly reported that 11% of IDPs left their homes due to water scarcity – exceeding conflict, security and livelihoods as a reason for migration (IOM, 2018). Incidents of tension over water access are now common throughout the country, reported in 38 different locations in Baghdad (UN, 2013). Water-related conflicts have been reported in Kirkuk, with farmers in Karbala abandoning their farms and livelihoods due to water scarcity (*ibid.*). Such events clearly illustrate one of the many ways water crises can impact vulnerable populations.

The Government of Iraq's Ministry of Planning outlined the main reasons for the country's water crisis in a UN 2010 Environmental Survey, as follows:

- inadequate management of water resources at all levels;
- large volumes of water required by construction projects;
- lack of a long-term strategic water infrastructure project; and
- lack of long-term, comprehensive strategy to combat the impact of water scarcity.

It is against this backdrop that effective solutions must be found to overcome three specific and significant barriers to sustainable water services in Iraq: dwindling water resources; the deterioration of the water and sanitation infrastructure over recent decades and; unstable access to power to properly operate the requisite systems.

To improve access to sustainable, resilient, affordable, equitable and safe water services, the Government of Iraq and the United Nations Children's Fund (UNICEF) developed solar-powered water supply systems for two vulnerable areas in rural Iraq: Shekhan, Ninewa, and Makhmur, Erbil. The programme was implemented in cooperation with the local government water service provider in the northern governorates.

Initially, water scarcity interventions had focussed simply on rehabilitating basic water systems and services to enhance access to sustainable drinking water services. However, due to lack of fuel and financial resources, maintaining the restored water systems proved challenging. To overcome this barrier, UNICEF worked with the Iraqi water administration to explore the potential of solar power as an alternative source of energy for water systems in the most vulnerable areas. The initiative to transition to a more sustainable source of energy was augmented by a range of other interventions to enhance resilience to water scarcity, such as water conservation behaviours and water efficiency. The programme offers a cost-effective solution to the provision of a sustainable, safe, low-carbon water services to vulnerable and displaced people.

This chapter examines the context for water resources in Iraq, considers the problems faced by the water sector and assesses the role of water authorities and institutions. It then describes the actions undertaken by UNICEF and the Iraqi water authorities in Shekhan and Makhmur, as well as the details of the solar-powered water system solution and goes on to analyse the overall impact of the programme.

8.2 CONTEXT
8.2.1 Water context in Iraq

The Tigris and Euphrates rivers are the main sources of water in Iraq, providing 93%–98% of Iraq's total water resources, while groundwater accounts for approximately 2%–7%. Other potential water resources such as desalinated water, treated wastewater and reused agriculture drainage water are still either underdeveloped or unused in Iraq (Abd-El-Mooty *et al.*, 2016).

Official figures from the Ministry of Water Resources (MoWR) indicate that the average annual flow of both the Tigris and Euphrates have decreased significantly over recent decades, with flows from the Tigris declining from 66 billion cubic metres (BCM) in the early 1970s to 28 BCM in 2015 and the Euphrates down from an average of 30 BCM in the 1970s to 8 BCM in 2015.

Figures from the Government's Water and Land Resources Strategy (MoWR, 2014) show that the agriculture sector requires over 40 BCM each year (see Figure 8.1), representing almost 60% of the entire national water requirement. Much of this consumption is wasted due to the inefficient methods used by farmers to irrigate their crops (Ministry of Water Resources, 2014).

If nothing is done to address the critical water situation and integrate resilience into water and sanitation services (UNDP, 2014), current projections indicate that the country's water deficit (the difference between the available water resources and the water needed) will increase from 11 BCM in 2015 to 29 BCM by 2035 – meaning an estimated 28% of the population will not have access to at least basic water and sanitation services (*ibid.*).

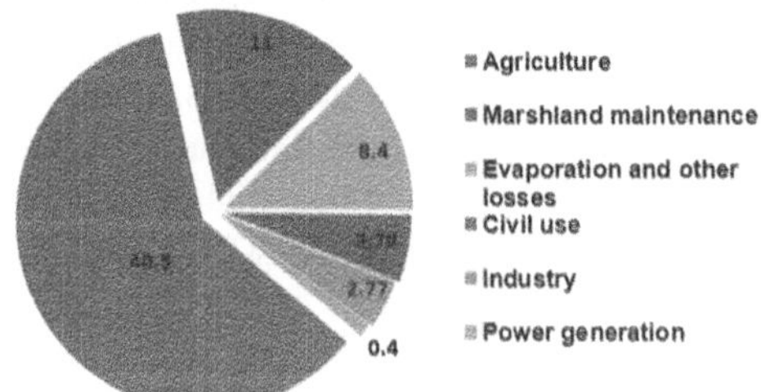

Figure 8.1 Iraq's annual water demand in BCM. (*Source*: Ministry of water resources, 2014).

8.2.2 The causes of increasing water scarcity in Iraq

There are a range of reasons for Iraq's water scarcity crisis, which affects the ability of providers to supply reliable water services. The elements described below are all factors causing or exacerbating the severity of the water crisis in Iraq.

8.2.2.1 Upstream riverine development

Iraq shares the water flow from the Tigris and Euphrates with Turkey, Syria and Iran, with only 8% of Iraq's total water supply coming from internal sources (Adamo *et al.*, 2018).

As a result of anticipated increased water demand in Turkey and Syria, it is likely that these countries will significantly increase their share of Tigris and Euphrates water resources in the coming years, drastically reducing river flows into Iraq (Lelieveld *et al.*, 2016; Tabari & Willems, 2018).

There has been significant upstream development on the Tigris and Euphrates in Turkey and Syria over the past 30 years, with construction of dozens of water-retention structures already completed and many others under construction, all having a direct impact upon water flows into Iraq (HRW, 2019a, 2019b, 2019c). Iran also has dam projects on the Tigris tributaries within its territory and there are dams in Syria on the Euphrates for hydroelectric power and irrigation (*ibid.*). A report by the International Association of Universities predicted a worst-case scenario of both the Tigris and Euphrates rivers being depleted by 2040, if no action is taken to secure a share of water for Iraq (IAU, 2011). In addition, precipitation in the region is expected to fall by up to 16% by the middle of this century (*ibid.*).

8.2.2.2 Climate change

In recent years, temperature increases and reduced rainfall have led to severe drought) in Iraq. Drought has affected both agriculture and water supply, as most Iraqi wheat production is rain-fed. The combination of reduced rainfall and extreme heat has led to an increase in evaporation and evapotranspiration rates, resulting in a large volume of water being lost every year, especially in southern Iraq (Frenken, 2009).

The extended drought conditions have had a detrimental impact on the lives of Iraqi citizens. The poor access to a reliable water service has led to a deterioration in livelihoods and crop production, an increase in unemployment and an increase in some diseases such as typhoid and diarrhoea (UNICEF, 2016).

8.2.2.3 Increased water consumption

The Iraqi population has grown from around 10 million in 1970 to almost 41 million in 2021 (worldometer, 2021). This population increase has been accompanied by a large number of refugees fleeing the conflict in Syria, as well as an increasing level of internally displaced people (IDPs), leading to concentrations of people and placing additional stress on existing services (IAU, 2011). To mitigate the impact of the reduced volumes of available water and the increased water demands, the Iraqi government has attempted to respond to reduce water consumption from 350 to 200 litres per capita per day (l/c/d), through a number of programmes aimed at raise awareness of water conservation and reuse (Ministry of Municipalities & Public Work, 2011).

Another factor which has contributed to poor water service delivery, in addition to the reduced volumes and increased water demand, has been an increase in competition for electricity, with demand rising by 24% in 2017 alone (World Bank, 2018).

In Iraq, 98.9% of electricity used is generated from burning fossil fuels, with oil representing 66.7% (413 terawatt-hours (TWH)) and gas, 32.1% (199 TWH) (Oxford University, 2019). The remainder includes water-based hydroelectric generation, representing 1% (6 TWH) and a small amount of solar generation, 0.2% (1 TWH) (*ibid.*). Oil and gas supplies are subject to significant levels of interruption, resulting in an unstable energy supply across most of the country (*ibid.*). While the Ministry of Electricity has built more plants to boost power production, this has increased water consumption as oil and gas power stations use steam in the generation process.

The demand for electricity, with its associated increase in water demand, is only likely to increase in the near future. Large numbers of industrial facilities, which have been out of operation (or near peak capacity) since 2003 or earlier, are expected to resume operations in the next few years, placing even greater demands on energy generation and therefore the water supply sector (World Bank, 2018).

A further reason for increased water demand in the country is the restoration of the marshes of Iraq, which were purposefully and extensively dried out under the earlier regime. These marshlands have been partially revived since the fall of the regime in 2003 and were named as a UNESCO World Heritage site in 2016. Work to restore the marshes continues, and it is believed 70%−75% will be restored within the next few years. The marshes require 15 BCM of water for restoration without improving water quality and an extra 5 BCM to improve water quality (World Bank, 2006).

8.2.2.4 Deteriorating critical infrastructure

In 2019, the United Nations' Humanitarian Needs Assessment found that 2.2 million people do not have access to safe and appropriate water and sanitation services in Iraq, of which 47% are children (Iraq Multiple Indicator Cluster Survey, 2018). However, even for those systems which are operational, most of the country's water facilities are working below the minimum operating standards of the World Health Organization (WHO) (*ibid.*).

The protracted conflict in Iraq has resulted in the destruction and extensive damage to water infrastructure, with decades of sanctions further compounding the situation. Cumulatively, these have severely impacted the expansion or rehabilitation/updating of water systems, as well as impeding the maintenance of the operational systems. The estimated damage to the Water, Sanitation and Health (WASH) sector in Iraq pre-2014 is US$1.4 billion (World Bank, 2018). Although there have been extensive rehabilitation efforts by the Iraqi Government and international development agencies, a large proportion of Iraq's infrastructure remains either damaged or destroyed, directly impacting the volume, quality and reliability of water and other services delivered to households (UNICEF, 2016). Extensive leaks occur throughout the water supply network and the volume and quality of freshwater has declined steadily over many years due to drought and river inflows. Inadequate treatment has led to sewage being discharged directly into rivers. The chronic shortage of electricity compromises water system infrastructure even further, with unstable power supply and frequent outages lowering water production and distribution (*ibid.*).

8.2.2.5 Lack of investment required to maintain and expand water services

The total capital investment required to connect all households to the public water supply network was estimated at US$6.6 billion for the period from 2016 to 2025 (UNICEF, 2016), with inflation causing the per capita cost to triple since 2000. Conflict has compounded the challenges to improve the levels of water and wastewater services, with ISIS and the ensuing economic crisis having caused water projects either to stop or slow down, severely impacting any progress that had been made (World Bank, 2018).

To repair and reconstruct water systems after damage caused by direct attacks to those systems as part of the conflict requires enormous amounts of funding and effort (*ibid.*). The investment needed to develop urban water systems in Iraq in the first place takes decades of concerted action that focuses not only on infrastructure but also on sector strengthening and institution building (*ibid.*). Protracted emergencies, such as that in Iraq, typically attract low levels of funding, with funding levels often insufficient to fully rebuild destroyed infrastructure. Iraq's government and the World Bank estimated Iraq's total recovery and reconstruction needs at US$88 billion, which included US$ 2442 million for the WASH sector, across the directly affected governorates over five years (*ibid.*).

8.2.2.6 Institutional failures

Institutional failures have led to inefficient water use and unreliable water services in Iraq. The failure to create incentives to manage demand and implement effective tariff modalities throughout Iraq illustrates how water is undervalued in the country. Iraq has one of the lowest water tariffs ($0.01 per cubic metre) in the MENA region and a high proportion of GDP is spent on public water subsidies (Human Rights Watch, 2019a, 2019b, 2019c; World Bank, 2018). With the ISIS conflict and economic crisis, accompanied by weak enforcement of water tariffs, revenues have fallen short of the levels required for infrastructure development and maintenance, including increased costs of abstraction and treatment. Most local and federal authorities have been reluctant to raise awareness of responsible water usage or the need to change people's water usage habits (World Bank, 2018).

Water services in Iraq are heavily centralised across governmental organisations. As there is no official water law, there is no official guidance to regulate the working mechanisms between these ministries. The segmentation of responsibilities for individual aspects of the water supply chain among five separate government departments and lack of coordination across government institutions has led to failures to address the complex water challenges in the country (UNICEF & UN HABITAT, 2013; Mumssen & Triche, 2017). Iraq's National Development Plan 2018–2022 (Ministry of Planning, 2018) has recommended establishing a 'National Water Council' to formalise the coordination mechanisms among relevant stakeholders in the water sector, in order to manage water resources at a national level and seek/oversee agreements at a transboundary level (*ibid.*).

Compounding the limited coordination amongst water sector institutions in Iraq is a lack of regulation relating to water supply. These two factors combine to impede the development and implementation of possible solutions for example smarter, targeted water tariffs or encouraging public–private partnerships, which could help build a more resilient water sector. Some of the main challenges facing the water sector in Iraq are outlined in Table 8.1, framed using IRC's 'building blocks' for a healthy WASH sector (Huston & Moriarty, 2018).

The role of the private sector in developing water and sanitation infrastructure, and in delivering and maintaining services, is limited in Iraq with investment in water services almost negligible. The private sector works with the local mayoralties to collect water tariffs and there are some private sector operations running small-to-medium-sized reverse osmosis (RO) water purification stations to supply water. However, the majority of these stations are unlicensed and unregulated and often the water quality does not comply with national drinking water standards.

8.3 THE SOLAR-POWERED WATER SYSTEM PROGRAMME

8.3.1 Programming for water scarcity in Iraq's WASH sector

The solar-powered water system programme was developed against a backdrop of broader strategic support which UNICEF has provided to Iraq's WASH sector.

Table 8.1 Challenges in Iraq's water sector.

Trend/Factor Impacting Adequate WASH Provision	WASH Sector 'Building Block' Impacted	Impact
Upstream riverine development (dams)	Water resource management	– Deterioration in water quantity downstream, largely in areas of southern Iraq
Climate change	Water resource management	– Reduced water availability (reduced rainfall and increased evapotranspiration/evaporation)
Increased water consumption	Planning	– Increasing deficit between available water resources and water demand
Deteriorating critical infrastructure	Infrastructure	– The extensive damage/destruction of WASH systems deeply affects their functionality in the cities and governorates, thereby affecting water services
Lack of proper funding	Finance	– Water projects are limited to operation and maintenance – Insufficient water and wastewater treatment – No new water projects are being planned or implemented – services are not extended or improved
Lack of long-term strategic plans and resilience approaches	Planning	– Failure to create incentives to address water scarcity across the country – Overlap of mandate and activities between ministries inhibits effective coordination
Poor water tariff structure and high subsidies	Policy and legislation	– Limited perception of the value of water, contributes to excessive and inefficient use – Limited revenue due to (a) the unwillingness of people to pay their water bills and (b) low tariff structure. Both of these inhibit the amount of maintenance which can be carried out, further compromising the quality and reliability of services
Limited role of private sector	Institutions	– Low investment and collaboration between the government and the private sector

Table 8.2 UNICEF-supported initiatives in Iraq's WASH.

Water Scarcity Responses	Detailed Interventions
Catchment water quality	• Strategy report on water quality management for Iraq
Demand management	• National Water Demand Management Plan, including water quality and wastewater treatment
Urban approaches	• Mobilisation on water conservation • Support to tariff frameworks and reducing non-revenue for water
Water supply enhancement	• Support to scale-up alternative technologies including reverse osmosis treatment, through public–private partnerships

UNICEF, alongside other sector partners, has worked with the Government of Iraq over a number of years to augment the water supply in the face of more complex and unpredictable threats through the development of strategic actions in sector planning and water management. This support has involved assisting government ministries to carry out assessments, identify and implement solutions and monitor their effectiveness. This approach has included considerations of catchment water quality, demand management, urban approaches and water supply enhancement, as set out in Table 8.2.

8.3.2 The intervention and local rationale

While UNICEF had already worked in cooperation with the local government water service provider in the northern governorates to rehabilitate the basic water systems and services in their coverage areas, challenges were faced to operate these rehabilitated systems. These challenges included unreliable power supplies (national grid and diesel generators), as well as a general lack of financial resources and qualified labour (Oxfam, 2018). In Iraq, water supply is generally powered by electricity from the local grid. The most vital link between critical water services relies on electrical power, provided by generating stations, distribution/transmission lines and substations, or, when a regular electricity supply is unavailable, on generators and fuel (*ibid.*; World Bank, 2018). A continuous electricity supply is required for all the components of the water supply systems to function; for example, water is moved throughout the distribution networks by pumped or gravity-fed systems, which requires electricity. Any attack on or shutdown of the electricity service (or generators and fuel) can disrupt water provision. Recent conflicts have seen many examples of attacks on elements of the electricity grid, leading to a loss of water supply. In addition, grid electrical power is becoming increasingly unreliable due to the heavy load placed on it, particularly in the summer months when air conditioning places a significant additional load on the system. The overall effect is to reduce the reliability and operational capacity of electrical supply and therefore water supply systems (*ibid.*).

Solar-powered water supply systems were introduced as a more sustainable, low carbon, alternative which required less maintenance and skilled labour and was a viable solution to increase the access to, and reliability of, safe water services for the most vulnerable and hard-to-reach communities of Shekhan and Makhmur, while reducing demand on the electricity grid.

The reasons for the rapid expansion of solar-powered water systems are many, but primarily relate to their rapidly reducing installation costs (and ever improving efficiency), lower maintenance costs, increased reliability and resilience, as well as lower carbon impact, particularly where fossil fuels are used to power water supply – as in the case with Iraq.

The solar-powered systems were installed in Shekhan and Makhmur, two rural districts in the north of Iraq. Both districts include Syrian refugee camps and the host communities have also been profoundly affected by the ongoing conflict and reduced access to water services.

Both districts suffer from water shortages, usually caused by ageing water infrastructure and limited access to the public water networks. Much of the water supplied to these areas is through private sector water trucks, which deliver to homes with capacity tanks on a weekly or monthly basis. The boreholes from which these trucks extract water in these districts contain high levels of sulphate and nitrate, often making water unsafe to drink.

The influx of refugees and IDPs has placed additional pressures on water services in both districts. UNICEF has worked with the government to ease this pressure, increasing the number of water trucks and supplying additional tanks to store water in the camps. Such solutions are often used at the onset of an emergency. The situation in Iraq is one of protracted emergency, with the lifespan of some camps exceeding five years. The design, installation and operation of solar-powered systems with reverse osmosis aims to provide a longer-term solution.

8.3.3 Advantages of solar-powered systems over diesel generators

Diesel generators are typically used in Iraq to operate water pumps in boreholes as their installation costs are usually cheaper than solar systems and generate similar amounts of power. However, in many rural areas it is difficult to maintain a regular supply of diesel and strict health and safety regulations must be implemented when diesel is stored in large volumes. In recent years in Iraq, diesel power is seen less and less as a sustainable option for the operation of water systems, particularly compared to solar-powered alternatives (see Table 8.3).

Table 8.3 Solar-powered system compared to diesel-powered system.

	Solar-Powered System	Diesel Generator
Maintenance	Less maintenance required Long-term warranties Most future maintenance (after the warranty expiry) can be done at a low cost	Regular maintenance required by skilled workers Regular changes of oil and filters are needed
Operation	An operator checks the system periodically, as it is automated	An operator must be available to operate and stop the generator and conduct basic checks
Safety	Few hazards if the system is designed and installed properly (a protective fence around the solar panels is recommended)	Higher potential for accidents, including fires (requires a protective shelter due to fire and electrical power risks)
Power generation	Variable, depending on the characteristics of the local weather conditions	Constant
Environmental	Low carbon impact	Generates emissions that exacerbate climate change and can be very noisy
Resilience	The system is fully operated by solar power and is less affected by interruptions for example conflicts, fuel/spare part interruption	The diesel-powered pumps will not work if diesel supply is interrupted due to interruptions to fuel and spare parts

Solar-powered water systems are low carbon solutions. They can reduce greenhouse gas (GHG) emissions when chosen over either diesel generators or grid-based electricity to power water systems, particularly in a country such as Iraq, where grid-based electricity is almost entirely generated using fossil fuels. Extensive use of solar-powered water systems could therefore lead to substantial reduction in GHG emissions.

8.3.4 How the solar-powered water systems work

The systems which were installed consist of a number of elements. A submersible pump abstracts water from the borehole and pumps to a reverse osmosis (RO) system. The RO system treats the water by applying pressure to the water on one side of a selective membrane. The impurities are retained on the pressurised side and the clean water passes through to the other side. The water is subsequently stored in an elevated tank and treated using a chlorine injection unit. All of the system's electrical parts (pumps, RO and chlorine unit) are powered by solar panels (Figure 8.2).

8.3.5 Outcomes

In Shekhan, 102 solar panels produce 32 kilowatts (kWh) of power (at 0.315 kWh each). The submersible pump power rate is 15–22 kWh. The discharge capacity of the borehole is 28 m^3/h. The average daily provision of water from the solar-powered water system in Shekhan is 250 m^3 and operates for 10 h in the summer without the need for supplemental power from the national grid or generator. On cloudy days during winter, the system produces 30% less energy on average. The systems have supplied at most, 20,000 people with reliable water supplies. They have increased access to safe drinking water, in excess of the minimum standards (100 litres per person per day), in these areas.

Figure 8.2 Solar-powered water project in Shekhan.

A comprehensive one-day training session has been provided to local government technicians on system maintenance and troubleshooting. The training was delivered immediately after completion of the system installation and was designed to support local capacity development and enhance closer coordination with the surrounding water directorate. UNICEF has also worked with the local water authorities to conduct water tests, to ensure the water is fully compliant with WHO drinking water quality standards. Furthermore, the project was handed over to the service provider at the local governorate who is responsible for operation and maintenance and all costs will be covered by the federal government financial budget location.

8.4 CHALLENGES AND OBSERVATIONS

While the project has been very successful in highlighting the potential of solar power for water supply services in Iraq, significant challenges were encountered.

A challenge was reallocating alternative land to implement the project due to local community and legal issues which became clear after the commencement of the project. Both challenges prolonged the project implementation period and highlighted the importance of clarifying the land ownership legal status in writing with all the relevant authorities and surrounding communities during the planning phase. In addition, it is important to clarify how national and/or local tax and customs regulation affects any necessary equipment importation well ahead of purchase.

UNICEF Iraq is planning to scale up the solar-powered system to further sites in rural areas in which disadvantaged populations are settled and are affected by variable quality of services. The aim will be to encourage sustainable and environmentally conscious power solutions in areas where the water authority has a limited capacity to operate water systems and where people cannot easily afford fuel to run generators.

UNICEF and the private sector will also make efforts to strengthen local markets and improve supply chains. Additionally, capacity building at all levels, particularly for local contractors and technicians, is vital to further professionalise the solar-powered water system management model.

After a year of operation, the systems have proved to be feasible and effective. There have been no shutdowns, resulting in considerable cost savings compared to diesel generators. The solar-powered water systems have provided a sustainable and reliable supply of safe water for hard-to-reach, vulnerable communities in Shekhan and Makhmur, while at the same time reducing air and noise pollution, emissions and operating costs.

The solar-powered water systems have produced sufficient water volumes and proven easier to operate and maintain than diesel-generator systems. This means they require less resources to keep them running and have the potential to offer a more sustainable solution for providing water systems to harder to reach, vulnerable communities.

8.5 CONCLUSIONS

The water crisis in Iraq is an urgent, complex, problem caused by numerous factors. It will require new, innovative, solutions and the development of water systems that are affordable, scalable, energy and water efficient and climate-smart. Solar-powered water systems have the potential to meet all of these criteria. However, solar systems, as with any motorised system, needs to ensure that current and future yields have been correctly estimated and that the systems are climate resilient and that water levels and quality are monitored.

To deliver solar systems successfully and sustainably across Iraq, further support will be required to strengthen the enabling environment (including oversight), build capacity, engage the private sector to strengthen local markets and improve supply chains.

Solar-powered water systems can be a more sustainable, reliable and low carbon way to supply water and can be a particularly effective solution for remote locations that are beyond the reach of the national grid network. Properly designed and maintained systems can help to alleviate the impacts of water scarcity in areas where access to water is a challenge and in regions where demand has risen quickly. It has also made an important switch from fossil fuels to renewable, clean energy sources has the potential to mitigate the negative effects of GHGs.

The results have shown that solar-powered water systems have performed well. They offer some potential to ensure greater resilience in the fragile, conflict-affected rural locations, where poverty levels are high and communities are greatly exposed to a range of threats, most significantly conflict and endemic water scarcity. The systems have ensured greater resilience in these communities, leaving them less dependent on fuel supply, which is frequently disrupted in Iraq. Despite this, investment in water services in rural Iraq is a long term, complex, endeavour. In this limited perspective, solar water systems may offer some resilience for the rural communities, where it is likely to take many decades for water services to have the necessary finances, human and technological capacity in place to deliver affordable and sustainable services for all.

The local governments' authorities at the governorates level need to be engaged to scale up such an initiative to provide water services to the most deprived population living in rural areas and suffer from unsustainable electrical power supply. Efforts should be made to ensure solar-powered technologies are made accessible and affordable for everyone. In terms of reaching the absolute poorest, this service delivery model should include the water price and include subsidies to support most marginalized households where required.

While solar-powered water systems require less resources to operate and maintain, it is nevertheless crucial to develop the capacities of both the public and private sectors, so that they can successfully install and manage systems, strengthen markets and improve accountability. Thus, UNICEF needs to continue advocating with the federal and local governments to invest more in solar-powered systems and build public and private sector capacities to successfully install and manage systems, strengthen markets, and improve accountability. This is imperative if solar-powered systems and other water supply systems fully meet the needs of the communities they serve. If these improvements can be made, solar technology can provide a significant opportunity to achieve universal and sustainable water access for millions of people.

Using the experiences and lessons of this programme as a foundation, there is no doubt that solar technology has the potential to be one of the cornerstones to achieving sustainable access to water services for millions of people in rural, water scarce, environments.

REFERENCES

Abd-El-Mooty M., Kansoh R. and Abdulhadi A. (2016). Challenges of water resources in Iraq. *Hydrology Current Research*, **7**(4), 1–8.

Adamo N., Al-Ansari N., Sissakian V. K., Knutsson S. and Laue J. (2018). Climate change: consequences on Iraq's environment. *Journal of Earth Sciences and Geotechnical Engineering*, **8**(3), 43–58.

Frenken K. (2009). Irrigation in the Middle East region in figures AQUASTAT survey-2008. *FAO Water Reports*, **34**, 402.

Human Rights Watch (2019a). Interview with Dr Shukri al-Hassan, marine science lecturer at Basra University, Basra, 16 January 2019.

Human Rights Watch (2019b). Interview with international engineering expert (name and location withheld), 7 February 2019.

Human Rights Watch (HRW) (2019c). Basra is Thirsty Iraq's Failure to Manage the Water Crisis. Human Right Watch Investigation, Basra. Available at https://www.hrw.org/report/2019/07/22/basra-thirsty/iraqs-failure-manage-water-crisis, 22 July 2019.

Huston A. and Moriarty P. (2018). Understanding the WASH system and its building blocks: building strong WASH systems for the SDGs. *IRC WASH: Delft, The Netherlands*, **40**, 17–18.

Inter-Agency Information and Analysis Unit (IAU) (2011). Water in Iraq Factsheet, March.

International Organization for Migration (IOM) (2018). No Water No Home: Assessing Iraq's Water Crisis Displacement Risk in Southern Iraq, Displacement Tracking Matrix'.

Iraq Multiple Indicator Cluster Survey Briefing (2018). Available at https://www.unicef.org/iraq/reports/2018-multiple-indicator-cluster-survey-mics6-briefing, 21 October 2019.

Lelieveld J., Proestos Y., Hadjinicolaou P., Tanarhte M., Tyrlis E. and Zittis G. (2016). Strongly increasing heat extremes in the Middle East and north Africa (MENA) in the 21st century. *Climatic Change*, **137**(1), 245–260.

Maddocks A. and Luo T. (2015). World Resources Institute Aqueduct Water Risk Atlas. Washington, DC, USA.

Ministry of Municipalities and Public Work-Iraq (2011). Water demand and supply in Iraq: Vision, Approach and Efforts. GD for water. Available at http://www.mmpw.gov.iq/, June 2011.

Ministry of Planning (2010a). Environmental Survey in Iraq, Water, Sanitation, Municipal Services, Republic of Iraq, Government of Iraq.

Ministry of Planning (2010b). National Development of Plan 2010–2014, Government of Iraq.

Ministry of Planning (2018). Strategy for the Reduction of Poverty in Iraq, 2018–2022, Ministry of Planning, The Republic of Iraq, Available at https://mop.gov. iq/en/static/uploads/1/pdf/15192838546d2344468c97dc099300d987509ebf27-- Summary.pdf, January 2018.

Ministry of Water Resources in Iraq (2014). Strategy for Water and Land Resources in Iraq, The National Centre for Water Resources Management-Ministry of Water Resources, The Republic of Iraq.

Mumssen Y. and Triche T. A. (2017). Status of water sector regulation in the Middle East and North Africa. *World Bank*.

Oxfam (2018). The Future of Humanitarian Water Provision is Solar, views-voices.oxfam. org.uk/2018/08/future-humanitarian-water-provision-solar/, 29 August 2018.

Oxford University (2019). Available at https://ourworldindata.org/energy/country/iraq, September 2020.

Tabari H. and Willems P. (2018). More prolonged droughts by the end of the century in the Middle East. *Environmental Research Letters*, **13**(10), 104005.

UNDP (United Nation Development Programme) (2014). Iraq National Framework for Integrated Drought Risk Management (DRM), a study by ELARD submitted to UNDP-Iraq.

UNICEF (2019). Iraq Monthly Humanitarian Situation Report. Available at https:// reliefweb.int/report/iraq/unicef-iraq-monthly-humanitarian-situation-report-january- february-2019, 12 April 2019.

UNICEF in Iraq (2019). Water Scarcity Case Study: Iraq. Internal report.

United Nations Children's Fund and United Nations Human Settlements Programme (UNICEF and UN HABITAT) (2013). *Iraq-* Public Sector Modernisation Programme, Water and Sanitation Sector, Functional Review Report, Baghdad.

United Nations (UN) in Iraq (2013). Water in Iraq Factsheet, UN Inter-agency factsheet.

World Bank (2006). Iraq: Country Water Resources, Assistance Strategy: Addressing Major Threats to People's Livelihoods, Report no. 36297-IQ, p. 97.

World Bank (2018). Iraq Reconstruction & Investment: Damage and Needs Assessment of Affected Governorates, [pdf] World Bank, Available at http://documents1.world bank.org/curated/en/600181520000498420/pdf/123631-REVISED-Iraq-Reconstruc tion-and-Investment-Part-2-Damage-and-Needs-Assessment-of-Affected-Governorat es.pdf (accessed November).

World Health Organization (WHO) and United Nations Children's Fund (2019a). Progress on Household Drinking Water, Sanitation and Hygiene. WHO, Geneva, Switzerland.

World Health Organization and United Nations Children's Fund (WHO and UNICEF) (2019b). Progress on Household Drinking Water, Sanitation and Hygiene 2000–2017: Special Focus on Inequalities. WHO, New York.

Worldometer (2021). Iraq Population, Available at https://www.worldometers.info/world- population/iraq-population/, 1 July 2021.

Chapter 9

Economic resilience in water supply service in rural Tajikistan: A case study from Oxfam

Orkhan Aliyev

WASH Programme Manager, Oxfam International, Dushanbe, Tajikistan

ABSTRACT

The water utilities established by the Tajikistan Water Supply and Sanitation (TajWSS) project, which is funded by Swiss Agency for Development and Cooperation and led by Oxfam in Tajikistan in collaboration with Government of Tajikistan focussed mainly on decentralization of drinking water services and ownership of the water supply assets by local governance bodies. However, owing to the increasing demand for water and pressures on water resources as a result of climatic variability, water utilities in rural areas are facing financial, operational and environmental challenges which prevent them from responding adequately. These challenges require highly resilient considerations in the design, construction and management of water supply and sanitation facilities and access to financial resources to overcome unforeseen risks. Oxfam's experience in Tajikistan shows that a community's socio-economic status and water utilities' business operations were key factors for building the resilience of water and sanitation (WS) systems in rural areas. In this paper, the approach in building WS systems that are highly resilient to disasters or risks in rural areas is investigated along with how different factors such as demand and supply, institutional capacity, access to finance and community ownership affect the sustainability of WS services.

Keywords: water supply, sanitation and hygiene promotion (WASH), economic resilience, WS system, water utilities

9.1 INTRODUCTION

Tajikistan is a small landlocked country in Central Asia, bordered by Afghanistan, China, the Kyrgyz Republic and Uzbekistan. The population of Tajikistan is 9 million (2018), with an annual population growth rate of 2.13%. Although Tajikistan's poverty rate declined from 47% in 2009 to 31% in 2017 (World Bank, 2017), its poverty remains the highest among the former Soviet countries and is concentrated mainly in rural areas where 76% of the poor live.

Tajikistan encompasses a territory of approximately 143,100 km^2, where 93% of this land is covered by mountains, glaciers and windswept plateaus. This means that only 7% of the territory is habitable, and only 5% of the land is arable. The country is heavily shaped by its physical geography, such as high mountain ranges, scarce arable land and uneven distribution of natural resources (especially water), which determines its agricultural practices, industrial development, transportation routes and infrastructure perspectives (Marveeva, 2009).

Tajikistan's economic development is quite complex, which is characterized by poverty, unemployment, underdeveloped real economic sectors and high remittance inflows caused by the outflow of labour migrants abroad (31% of GDP in 2018) (Tajikistan-World Bank Group Country Partnership Framework 2019–2023, 2019; World Bank, 2019). The agricultural sector is the largest employer (particularly in rural areas), although it provides the lowest wages. Deteriorating infrastructure and services, degradation caused by over-exploitation of arable lands and water resources, use of outdated technologies and other factors limit opportunities for economic diversification and employment opportunities. Tajikistan's social characteristics are primarily driven by its rapidly growing population, particularly in rural areas. Tajikistan has the youngest population in Central Asia (35% of the population is between 14 and 30 years of age). However, more than one in three young people – including nine in ten young women – are either unemployed or not pursuing education (World Bank, 2019) (*ibid.*). Despite increasing budget allocations towards social protection, the overall health status of the population remains poor.

9.2 ACCESS TO DRINKING WATER AND WATER GOVERNANCE COMPLEXITY

9.2.1 Water resources

The glaciers and snowfields of Tajikistan are the main water towers for the whole region in Central Asia and are thus critical for economic development of downstream countries. Those glaciers contribute 10–20% to the total runoff of all major rivers in the region, which make up 40–60% of all water resources in Central Asia (Mustaeva *et al.*, 2015). Being water-rich with huge reserves of hydrological resources, Tajikistan has strong hydropower potential from extensive mountain and water resources, but it uses only 5% of its hydropower potential, which is the source of over 90% of its electricity (*ibid.*, p. 12). Therefore, water resources play a critical role in the economic development of Tajikistan, especially for agriculture and energy. Moreover, its dependence on hydropower makes the country vulnerable to fluctuations in rainfall and climate change, which have adverse effects on energy and food security, poverty and human health.

9.2.2 Access to water supply and sanitation (WSS) services

Since the collapse of the Soviet Union, there has been very little investment into basic infrastructure and social services in Tajikistan. Moreover, the civil war in Tajikistan from 1992 to 1997 left the economy in ruins. This has significantly delayed the establishment of public services and investment in the country, which culminated in adverse effects on people's economic and social lives. Most of the WS and sewerage systems built during the Soviet period have rapidly deteriorated across the country, and lack of public maintenance and investment has led to serious infrastructure degradation and service deficiencies.

Although Tajikistan enjoys abundant freshwater, access to improved drinking water sources and sewerage systems remains significantly low compared with other Central Asian countries. Rural areas are particularly badly affected, and progress in achieving sustainable WS services has been frustratingly slow for the rural population.

As of 2017, only 48% of the population of Tajikistan have access to a safely managed drinking water service, but access to basic water services in rural areas in Tajikistan is 76% (Figure 9.1). This is mainly driven by the replacement of surface water with water from public standpipes and 'the neighbour's tap'. Figure 9.1 also illustrates the fact that rural areas are predominantly devoid of service providers. Sanitation is a particularly neglected area in Tajikistan. Only 18.2% of the population in small towns has access to sewage systems, while this

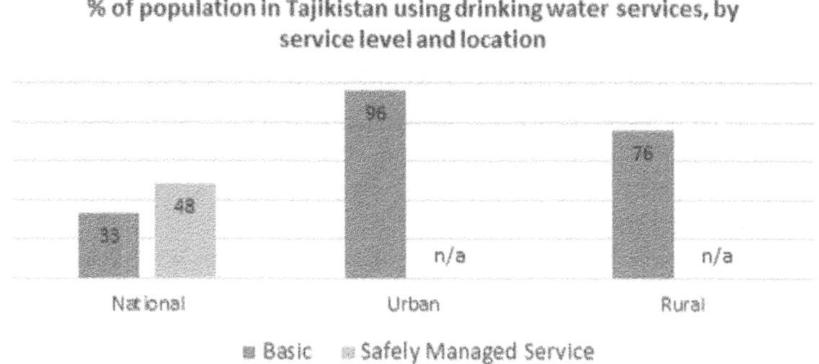

Figure 9.1 UN Water, SDG 6 in Tajikistan (2017) (UN Water, 2021).

is represented by only 0.2% of the rural population (World Bank, 2017). Many are reliant on pit latrines in peri-urban and rural areas, and the wastewater produced by households and industries is discharged into the soil and environment without treatment.

9.2.3 Water governance issue

The key drivers of change for water supply, sanitation and hygiene promotion (WASH) sector reform in Tajikistan are population growth, increased water demands from agriculture and industry, climatic variability and climate change. These growing pressures mean that Tajikistan must continue to invest and develop its water, sanitation and hygiene services for social and economic growth. However, the institutional structure of the water and sanitation (WS) sector is so complex and fragmented that the establishment of service provision poses significant bureaucratic challenges to the sector development. The State Unitary Enterprise (SUE KMK) is both the regulator and operator of drinking water and sanitation services in Tajikistan, but the Ministry of Energy and Water Resources (MEWR) acts as the water sector policy regulator. There is a huge uncertainty regarding the relationship between these two state bodies, which has exacerbated the problem even further in rural areas with insufficient capital investments. This shows that the low access to drinking water and sanitation is clearly not due to absolute water scarcity but to the lack of good governance, contradictory legislation and blurred state responsibilities.

Since most of the rural areas are deprived of water services, the WS service delivery in those geographical areas is being implemented by a 'TojikObiDekhot' (a rural subsidiary of SUE KMK). However, most TojikObiDekhots are located in district centres and serve mostly peri-urban areas, if there is an existing WS system. If there is no existing WS system, TojikObiDekhots rarely provide any service, let alone investment. Moreover, given the mountainous landscape and challenging accessibility to remote villages, the state subsidiary organizations are not in a financial and operational position to provide services, which means that water services are not within the reach of the population who were accustomed to getting those services from the government during the Soviet period. However, people still hoped that those services would be established by the government, although their hopes have waned through generations. For that reason, the communities have nurtured a feeling of non-ownership over their assets and do not possess a shared responsibility (or feeling) for any assets or infrastructure ever built in their areas.

At the global level, Tajikistan is a member of the High-level Panel on Water launched by the World Bank and the UN, and, on the initiative of the President of the Republic of Tajikistan, the UN General Assembly adopted a resolution titled International Decade for Action 'Water for Sustainable Development' (2018–2028), where the government has reiterated its commitment to the availability and sustainable management of WS for all as part of the UN's Sustainable Development Goals, that is SDG 6, by 2030.

9.2.4 Market challenges and local realities

In Tajikistan, water resources belong to the state, and the law on 'Drinking Water and Wastewater' (Law of the Republic of Tajikistan N1633 2019) mandates that the government is the guarantor of access to drinking water for the population and institutions. A separate government resolution 'On Approval of the Procedure of State Control and Supervision of Drinking Water' (Resolution # 679 of the Government of Republic of Tajikistan, 2011) stipulates that SUE KMK is the agency responsible for the management of drinking water and wastewater treatment. It clearly states that KMK is the implementer of state policies and MEWR is the policy watchdog to enforce control and regulation over the implementing bodies; however, in reality, the exact boundary of responsibilities between MEWR and SUE KMK has not been precisely outlined to mobilize state funding appropriately.

Despite this, the WS sector in Tajikistan is characterized by an uneven distribution of state roles at the national, regional and district levels. For rural areas in particular, a WASH strategy appears to be non-existent, and no institution seems to be clearly responsible for such challenges. Ultimately, all WASH-related responsibilities are given to the local government (Hukumat, Jamoat), but the local government institutions lack sufficient financial and institutional resources for reform implementation.

Developing and expanding market-based WASH products and services in this environment are the other challenges for donors and implementers. However, the starting point in developing the WASH market is to create that demand and provide necessary support to service providers to meet the growing demand for WS products.

9.3 MARKET-BASED RESPONSES TO WATER CRISES IN TAJIKISTAN

9.3.1 Transition from humanitarian to development aid

To address this humanitarian crisis, Oxfam initiated operations in Tajikistan in 2001 with the humanitarian mission to alleviate human suffering and respond to natural disasters, especially in rural areas. In the WASH sector, Oxfam diverted its resources to address the rehabilitation of Soviet-built drinking water and sanitation systems, including the construction of public taps in rural areas and latrines in schools and healthcare facilities, that had deteriorated substantially over time. The approach was mainly supply-driven with no payment mechanism from households, given the fact that the rural inhabitants were considerably poor at that time and unable to afford payment.

Therefore, Oxfam in partnership with the Government of Tajikistan (GoT) conducted a study in 2007 on the status and performance of water utilities in rural areas. The research outcomes concluded that the status of access to drinking water for the rural population is very poor and there is very limited institutional capacity to address the issue at national level. Investment and technical know-how were two of the issues most commonly flagged up by local government representatives and communities.

It was identified that the centralized water supply systems at district or town level were in poor condition and often do not reach the villages or the most remote populations. Moreover, the quality and quantity of water provided at district level were often unreliable due to poor operation and maintenance.

To address this issue, Oxfam and the GoT held a series of consultations with key donors, civil society organizations and community representatives to improve basic water supply in rural areas. As a result of recommendations, in 2008, Oxfam began to focus on transitioning to a more sustainability led WASH programme approach with an emphasis on decentralized water supply services, public and private partnerships and policy advocacy.

The new sustainability led approach to WASH service signified a fundamental change in service delivery, where the centralized supply system was not feasible due to technical, economical and/or institutional reasons. The decentralized nature of the system referred to the distribution and treatment of water to the community, where the service is operated and maintained by community-level water user associations (WUAs). It was also expected that the decentralized infrastructure would enhance resilience by contributing to water resource conservation, cost efficiency and greater adaptability to configure water systems for the specific local contexts, and capacity of local operators.

9.3.2 Economic resilience approach

Natural disasters, which Tajikistan is very prone to, along with climate change and rapid urbanization pose a serious risk to rural areas, which are less financially viable for service providers in the drinking water and wastewater treatment sector. The WS systems in rural areas, if they exist at all, are particularly susceptible to risks from natural hazards and disasters. Moreover, given the level of impoverishment in rural areas, the government response to ensure a basic WS service becomes an issue owing to its limited capacity.

History. Oxfam refers to 'resilience' in Tajikistan as economic resilience in order to link the economic capacity of the community (demand side) with the service provider (supply side) to overcome the environmental, social and economic challenges. In this regard, the definition adopted by the United Nations Disaster Risk Reduction (UNDRR) office is more relevant as it describes the resilience as 'the ability of people, organisations and systems, using skills and resources, to manage adverse conditions, risk and disasters. The capacity to cope requires continuing awareness, resources and good management, both in normal times as well as during the disasters or adverse conditions' (Assembly UG. 2016).

Given that most resilience indicators focus on technical dimensions, Oxfam has transformed its WASH programme since 2009 to address the economic and governance dimensions of the WSS sector. The programme developed a strategic vision until 2022 and is based on a theory of change that supports the development of autonomous WASH management structures in the country to establish financially and operationally sustainable WS and wastewater treatment services, especially in rural areas.

The financial and operational sustainability of WS refers to recovery of operations and maintenance (O&M) costs intended to achieve a fully functioning water system to ensure the capture, treatment-purification, transport and active supply of water to consumers. Oxfam developed a plan for a cost-recovery tariff that can be used by WUAs. It is based on recovering reoccurring O&M costs, giving operators the ability to sell their services in exchange for water consumption.

Oxfam's cost-recovery tariff setting plan provides water utilities with financial means to cover their O&M expenses to reach a certain level of financial independence. The O&M services include but are not limited to:

Operation:

- Supervision and monitoring of equipment, machinery and other constituent parts
- Operation and management of technical elements, machinery and distribution intervals
- Process control (flow rates, sampling etc.)
- Consumption management (electricity tariff optimization)
- Risk and administrative management (risk assessment, mitigation plan, invoicing etc.)
- Waste management
- Reports

Maintenance:

- Electromechanical maintenance (equipment, repair, replacement etc.)
- Regulatory maintenance
- Upkeep (painting, leaks, carpentry, gardening, cleaning etc.)

Strategy. The programme is built upon four pillars (Table 9.1) that act as the transition into the market-based WASH service delivery and incentivizes business models to sustain economically viable service delivery in Tajikistan.

An interaction between these pillars is built by influencing policies and system changes based on best practices in the field so that the government can adopt adequate measures in water sector reform. Moreover, the market interaction between the service providers and consumers is the interaction with the most

Table 9.1 WASH programme strategy pillars.

Theory of change: Improved, sustainable and equitable access to safe drinking WS, and adoption of better hygiene behaviour in Tajikistan through market players			
Building demand for and supply of WASH services	Improved governance and resilience of WASH service providers	Securing financing and system functionality through blended funding	Policy advocacy and influencing
• Consumers pay for hardware and services • Utilities are sustained by consumer payments and public subsidy • Village is provided with regular electricity and fuel	• Water utilities are trained about effective management and technical monitoring approaches to address consumer needs and pre-empt operational challenges • Utilities provide services and sell products to consumers	• Blended financing of WASH infrastructures • Local government provides subsidies to water utilities	• Policy changes in legislation, regulation and guidance • Coordination of water sector reform

crucial role in people's behaviour, as it can respond quickly to people's needs when demand grows and offers a solution to consumers to contribute financially by buying its services or products. Moreover, the choice of the village also depends on the local context (the level of demand) and availability of public utilities, primarily regular electricity supply.

Oxfam in Tajikistan invested in the development and creation of small water utilities (community-based water user associations or state-owned water enterprises), and market-based solutions such as working with sanitation product retailers (latrines, toilets and hygiene products for households) and wastewater treatment systems (decentralized wastewater treatment systems and faecal sludge treatment plants). All these entrepreneurial type activities rely on consumer payments. This is a key element in developing and sustaining the service delivery with some degree of dependence on external funding, which builds the economic resilience of the water utilities to be robust and resourceful enough to withstand disasters and be able to return to a new normal after a disaster.

Thus, Oxfam predominantly utilizes the following combination in the project design and implementation (Table 9.2):

A further element that cuts across these three areas is to guide and facilitate the process for innovative products or services by market players, which can then be converted into a scalable business.

9.3.3 Community managed services to meet user demands

In the early stages of the feasibility study on site and with the community, it was revealed that the rural communities wanted what they were mostly deprived of – safe drinking water that is available when needed and in reasonable proximity. Most people, particularly women, indicated the long distance to fetch water, unreliable water quality with seasonal changes and lack of governmental support to fix problems on site. Though the water was mostly available in large quantities, the communities did not have the knowledge, capacity, and financial means to make accessibility easier. They had no belief in governmental support

Table 9.2 Market-based WASH programme design.

Demand creation	• Stimulate demand among people to the point of making a purchase
Supply development	• Build the capacity of water utilities to meet the demand created
Enabling environment	• Collaborate with the government and international stakeholders to revise policies

or any investment projects and relied mostly on the women's ability to reach the water. Interestingly, when questioned about the price tag for water supply, the majority of community members responded favourably to the idea of payment if the quality and availability of water were in place.

When Oxfam and local government compiled the survey and interview results from community members, the following list of demands was discovered (Table 9.3):

Table 9.3 Community demand.

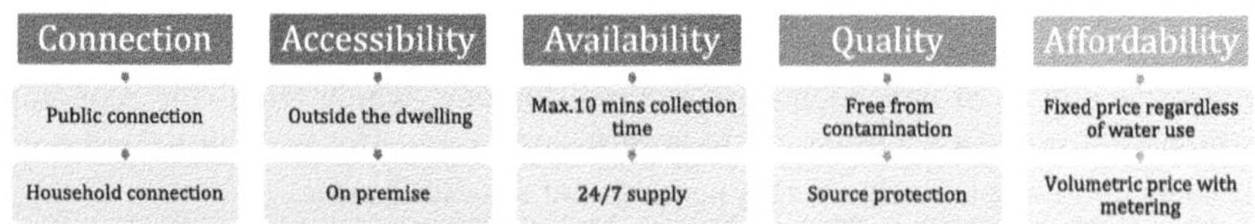

9.3.4 Water management model

After the community demand analysis for water, Oxfam together with the Tajik government and civil society organizations (CSOs) agreed to design a decentralized WS system with cost-recovery for communities. In the case of Tajikistan, given that the cost to connect villages to a centralized WS system is expensive, the decentralized WS system is considered more affordable. And there was a general consensus among stakeholders that the decentralized WS system would lead to large improvements in public health by making water available, reliable and safe to drink in areas where the centralized supply fails to provide it.

Ultimately, a water utility management model was designed that encapsulated all essential factors regarding the selection of an appropriate WS system. As the next step, a community-based water user association (WUA) was established within the local government's unit (Jamoat). The WS system as an asset belongs to the local government but the management duty is handed over to the WUA. The local government acts as the supervisor to regulate and monitor the WUA's performance, while the WUA is tasked with day-to-day management, service delivery and technical maintenance of the system. In case of capital re-investment, the government is engaged to finance the restoration of the functions of the WS system in the village.

Table 9.4 showcases the model. As seen from the table, the decentralized WS system requires daily operation and maintenance for users, therefore, decentralized supply has the benefit of putting users in control of their system maintenance.

Table 9.4 Water utility management model.

Source of Water	Water Purification Technology	Operation & Maintenance	Payment Method	Utility Dependence
Groundwater or Spring • Drilled wells • Pumped reservoir • Piping for gravity flow	• Chlorination • UV lamps	• Community-based water user associations (WUA) • Tojik Ob Dekhot (State Unitary Enterprise responsible for rural water supply • Limited Liability Company • Note: There are different legal structures established in Tajikistan, but in practice, they perform similarly in terms of O&M	• Private metering • Tariff allocation for consumption of 1 m^3 • Monthly household payment based on agreed tariff • State subsidy	• Fuel (service delivery) • Electricity (pumping, UV lamps, delivery etc.)

9.4 KEY LEARNINGS AND CHALLENGES IN BUILDING ECONOMIC RELIANCE OF WS SERVICES

9.4.1 Community resilience

Oxfam's past experience with rural WS service providers demonstrates that the primary economic factor affecting WS resilience is the economic capacity of communities. Moreover, owing to underinvestment, slow pace of reform and low interest of the private sector to invest in WS services, the communities have little expectation that the government will resolve this issue soon.

The lack or poor condition of infrastructure stimulates more demand for water because the cost of inadequate hygiene practices and poor sanitary conditions in schools, healthcare facilities and households in previous years outruns the cost of investment for today. Having studied this trend among the population in rural areas, the intervention approach has been diverted from public connections to private (household) ones to change the communities' experience with water.

It became clear during the implementation of the WASH programme in Tajikistan since 2009, that communities who were more engaged in the decision-making process have demonstrated ownership of the management structures in place and a willingness to contribute towards the facility's sustainability, thereby enabling a more resilient water service.

9.4.2 Institutional resilience

Lessons learnt from other projects and disasters have led to more proactive approaches to address the consequences of disasters. The government of Tajikistan mainly relies on aid from international donors to eliminate these risks. However, the starting response and restoration activities are usually subject to budgetary approval by the central government, which can be cumbersome and lengthy. Quick access to funds is very important in the first few weeks following a disaster. Ideally, the district government or local government structures should have different ex-ante (proactive) and ex-post (reactive) financing instruments to give water utilities quick access to finance and thus speed up activities in response to disasters. However, this is not the case.

Thus, Oxfam designs its projects to deliver capacity-building sessions for WS service providers from the very beginning of construction work and trains the community in ownership and self-financing instruments to protect their WS infrastructure. Water utilities can pay the response and restoration costs from their budget contingencies and/or reserves which may include communities' financial contributions. Experience shows that using contingencies and reserves is the quickest way to access finance in the aftermath of a disaster. However, contingencies and reserves are usually small and can quickly be depleted.

Depending on the severity of a disaster, any quick access to finance, even in small quantities, is better than long delays. Governmental support or funding from international donors typically takes time and requires the involvement of significant bureaucracy. The funding, if received, can be allocated for post-disaster recovery or construction activities that were not duly accomplished in the first response. Practice has suggested that even a small contingency budget is the safest option for a quick response to disasters to at least prevent the environmental risk from becoming colossal in the affected villages.

9.4.3 Economic resilience

To summarize the above-mentioned findings regarding community and institutional resilience, monitoring and evaluation results showcase that trained and fully informed communities are better at mobilizing against mitigation and/or response activities to protect WS systems. This, in turn, makes the WS systems of communities more resilient to disasters. Once the community understands that it is more cost-effective to invest in building more resilient systems with safety measures in the early stages of the construction phase, they are more prone to contribute earlier. On the other hand, water utilities' quick access to finance is also a significant factor in disaster risk reduction (DRR). However, funding availability does not necessarily guarantee a quick recovery if the community's support is not in place.

It is highly recommended that the water utilities in rural areas are duly and regularly trained on how to prepare for a disaster by setting up an accountable contingency plan and budget. In the absence of immediate external funding after a disaster, water utilities should be able to access funds from their existing budget by cutting unnecessary and non-urgent expenses. Preliminary budget estimation for contingency preparedness is crucial either by community fund raising or reducing capital expenditure by putting development works on hold. While the resilience of the WS system heavily depends on communities' economic capacity and water utilities' access to finance, other dimensions such as transparency and accountability in operational and financial work cannot be ignored.

9.5 DECENTRALIZED WATER GOVERNANCE AS MEANS TO BUILD STRONG RESILIENCE TO RISKS

In this section, experiences with implementing the business model for building economic resilience of WASH systems will be described. This section will also give an insight into how the market-based WASH programme can be designed and implemented considering the demand and supply sides. The market-based WASH programme was implemented with a governance model where all stakeholders are incentivized to invest first before any action is taken.

Table 9.5 shows the overall governance stages in market-based WASH programming.

District-level governance. The WS infrastructure project is tendered with the announcement of an investment plan in rural areas through district governments. The announcement specifically mentions Oxfam's conditional funding requirements in the WS system and the requirement from the community to contribute a minimum of 5% of the overall infrastructure cost. Once applications are received from rural municipalities, Oxfam and the district government shortlist the villages and carry out initial technical cost estimation, a community willingness survey and DRR assessment. After careful analyses, the selected village is notified about the result and a Memorandum of Understanding is signed between Oxfam and the district government, as well as the central

Table 9.5 Inclusive WASH programming through four stages.

District-level Governance	Service-level Governance	Community-level	Policy Advocacy
Water Trust Fund (site selection and investment)	Infrastructure building	Active citizenship	Partnership and Cooperation
Tariff and taxation	Operation and Maintenance (O&M)	Social accountability	Expert level discussion
Audit and evaluation	Value chain development	Hygiene education	Resource mobilization
Scaling and replication	Scaling and replication	Improvement of health status	Water sector reform

government (SUE KMK) on cost sharing at 10% and 15%, respectively. After the technical design and cost calculation for the WS system are complete, the final cost is assessed by Oxfam and district government engineers for fine-tuning. Once finalized, the project design, cost and management details are discussed with the central government, district government and communities.

Overall, the following key issues are verified and contractually agreed before the project starts:

- Co-financing arrangement – 70% Oxfam, 15% central government, 10% district government, 5% community
- Identification of water management body – Public or private, or community-based water user association
- Connection type – WS connection at household level
- Payment condition – Volumetric tariff system with meters installed in each household

Oxfam and the district government convene Water Trust Fund meetings – Governing Body for decision-making – where they agree on investment lines and funding delivery means. After selecting the construction company through bidding, Oxfam and the district government sign a separate contract for the building of the WS system. The communities either pay in cash to the construction company or contribute in-kind (labour force) for the designated amount of work. This way, all parties, from the very beginning of the project, become shareholders and establish a solid ownership over the assets and further processes. At this stage, the business plan is also discussed with the water utility where Oxfam and the government facilitate the design and possible funding and expansion plan.

Service-level governance. At the service level, Oxfam supports the creation or development of the WS service in rural areas. Initially, the new water utility is established and trained in technical and financial management aspects of the infrastructure work and WS system. Once it is established, the construction of the WS system is tendered and the newly established water utility is tasked with regular monitoring of the construction work on site. Before the completion of the construction stage, the water utility is involved in a series of training on tariff setting, technical and operational maintenance, taxation and accounting, customer data base collection and update, social accountability tools and communication.

Community-level governance. Active citizens' engagement is rigorously promoted alongside women's empowerment in target communities as water is very much a driver of economic development. Ironically, efforts to increase access to improved WASH services at the household level often do not adequately consider the risk related to public health. Moreover, in community-level awareness sessions, health-related costs that are often given little weight are highlighted. The social accountability dimension is established by a community advisory board (CAB) within the water utility, whose members are influential people from the community, to hold the water utility accountable for its operation, financial accounting and expansion plans. This CAB convenes once every quarter to listen to water utilities' reports and issues that can be solved through community mobilization.

Oxfam usually encourages community engagement in WASH-related activities to trigger consumer responsiveness towards service quality before, during and after a disaster period. It is very important to prepare the communities to handle the risks very quickly and responsibly using internal resources first in order to prevent the disaster from spiralling out of control. As a hazard, such as a flood or rockfall, can lead to a range of secondary hazards, people might be exposed to contaminated water in the WASH system.

Policy advocacy.

The WS sector in Tajikistan had an acute need for an arena where stakeholders could meet and share experience, ideas, views, knowledge and particular experiences related to the WASH (Water, Sanitation and Hygiene) sector. To respond to this need, Oxfam, in consultation with the government and the Swiss Agency for Development and Cooperation (SDC), as well as other stakeholders, initiated the Network of Stakeholders on Sustainable Water Supply and Sanitation (TajWSS Network) that was launched in November 2009. The network's financial support was provided by the SDC and facilitation was taken forward by Oxfam as an implementer of the SDC-funded Tajikistan Water Supply and Sanitation (TajWSS) project.

The network is now represented by more than 70 stakeholder organizations from the government, parliament, the UN, donors, academia, international non-governmental organizations (INGOs), civil society, the private sector and the media. Its goal is to advocate for policy reforms in the WASH sector and provide expert level support to the government based on the lessons learnt from the field, that align with the SDG 6 targets.

9.6 MAJOR CHALLENGES IN ESTABLISHING RESILIENCE OF WS SYSTEM

Most rural water utilities who were interviewed (Oxfam, 2021) mentioned five key challenges that pose risks to the functionality of the WS system. A limited review of water user associations (WUAs) highlighted the following as risks to their business:

- Higher operational cost
- Operational malfunction
- High employee turnover rate
- Illegal connection by communities
- Limited support by the government and community

High operational costs. The operational cost of WS systems increases with every passing year (Figure 9.2). Thus, the water utilities try to save money while

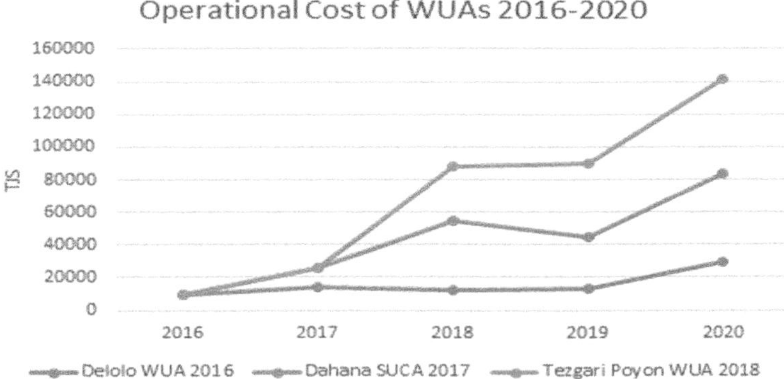

Figure 9.2 Operational cost of WUAs (Oxfam Management Information System (MIS), 2021).

running the system and/or negotiate with the communities to increase the tariff in order to cover the cost. Given unstable water supply means because of climate change and the inflation rate, the operational and maintenance cost (O&M) of the WS system increases every year leading to rises in tariff.

Over the period from 2016 to 2020, the operational cost increased 206% in Delolo WUA in Muminabad district, 349% in Dahana WUA in Kulob district and 74% in Tezgari Poyon WUA in Rudaki district. This clearly demonstrates the water supply infrastructures are typically capital intensive and require a high sunk cost for longer operational functions. The WUAs can increase the operational efficiency through regular maintenance control and a risk register for communication with the customers, who could potentially contribute to the mitigation activities. It also illustrates that the tariffs for water services can generate only a share of the revenues needed, and a government subsidy is required to provide an appropriate level of service and mitigate the risks associated with the cost.

Operational malfunction. The WUAs have mentioned four main areas where an engineering intervention is inevitable, namely, water leakage from pipes, disruption of water meters (especially in winter), overconsumption through illegal connection and pump malfunction. These technical problems demand specialized expertise that is lacking in the villages, and water utilities are obliged to seek paid labour from neighbouring cities.

High employee turnover rate. The WUAs have seen the substantial risks caused by the departure of technical or financial staff for better jobs in the cities. Moreover, sometimes the district government unofficially dismisses the WUA chairman and replaces him with his subordinate, which is regarded as an abuse of power. This causes a severe disruption in the system and requires additional training and induction. Normally, the chairman of a WUA is selected by the community, and if the person is not respected or recognized by the community, a power struggle occurs.

Illegal water connection. The WUAs have detected more than 50 illegal water connections, either by households or neighbouring villages, to avoid payment. This causes a conflict in the community due to rapid consumption of water from the tanks that goes unnoticed by the water utilities.

Limited support from the community and government. Unfortunately, in all target districts, there is no subsidy scheme in place by the government to support the rural water utilities. Besides, the WUAs have also complained that the community members become less supportive when there is an interruption in the water delivery due to the adverse impact of environmental changes for example flooding, landslide, rockfall, pipe breaks or pump malfunction.

9.7 MAJOR OPPORTUNITIES IN ESTABLISHING RESILIENCE OF WS SYSTEM

As described above, WUAs have come across many challenges in managing the WS system. However, as small rural entities, they also use business opportunities in this work in the background of population growth and potential for expansion to neighbouring villages. During interviews, WUAs mentioned four key areas to improve that they think could be a good set of circumstances to grow as a social enterprise:

- Managerial skills based on key performance indicators (KPIs)
- Possibility to expand
- Tariff setting
- Social accountability

Managerial skills based on KPIs. Oxfam, in partnership with local non-governmental organizations, set up KPIs for rural water utilities to measure their financial and operational progress. Those indicators are shown in Table 9.6.

The KPI-driven water management from 2019 onwards incentivized the water utilities to perform better for higher profit and better customer support. Figure 9.3 also shows that the gross profit margin has started to grow since then. Given that all households are connected to water meters, the WUAs monitor them to detect leaks and eliminate wasteful uses, as well as ensuring sufficient drinking water in the reservoir and that adequate sanitation and hygiene behaviours are practised by communities. This serves as a success indicator for water utilities to measure their progress.

Possibility to expand. WUAs have received multiple requests to expand their household connections, which they see as an opportunity to obtain more profit in the long run. Oxfam also provides a consultancy support to water utilities on revenue generation as part of the business. As of 2020, most water utilities expanded on their expense with some financial contribution from potential

Table 9.6 KPIs for rural water utilities in 2018.

Water Utilities	Tariff Collection Ratio (% of bills)	Gross Profit Margin[a]	Customer Satisfaction Rate (%)	No. of Expansion Requests Sponsored (%)
Delolo WUA	71%	0.31	75	71
Dahana SUCE	82	0.10	77	100
Tezgari Poyon WUA	94	0.30	86	100
Average	80%	0.24	79	90

[a]Total operational revenues/total operational expenditures. If the figure is above 1, it indicates cost recovery.

customers. Table 9.7 perfectly illustrates the additional household requests and percentage of coverage expense paid by each water utility.

Tariff setting. Under Soviet times, citizens did not pay for water services, and tariffs have historically been set below the level at which service providers can conduct basic operation and maintenance thus leading to underperformance and a dependence on external funding. In Tajikistan, consumers pay too little for water services and the revenue from water charges does not even cover the operation and maintenance cost, let alone reinvestment for the infrastructure. Often, consumers are not aware of the real costs associated with the water supply services because these have been historically heavily subsidized by the government.

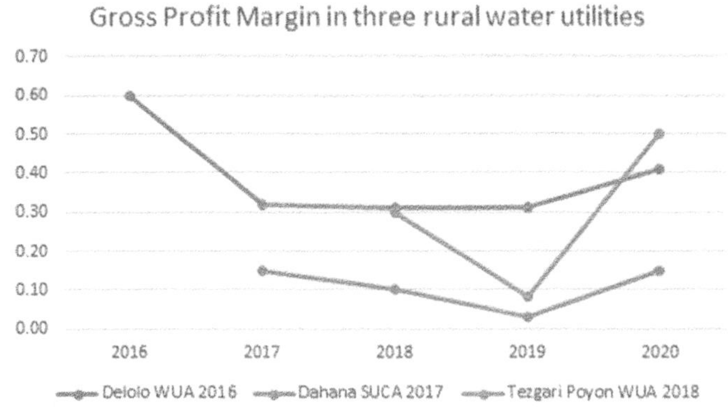

Figure 9.3 Gross profit margin (Oxfam Management Information System (MIS), 2021).

Table 9.7 WUAs' expansion progress.

Names of WUAs	No. of Requests Received	Expanded on WUA's Expense (%)
Delolo WUA 2016	112	71
Dahana WUA 2017	31	100
Tezgari Poyon WUA 2018	54	100

Oxfam employs a cost-recovery tariff methodology that is set by the Anti-Monopoly Agency of the Government of Tajikistan within the legal framework. The tariff setting process and decision-making are organized through involvement of community members to identify the local needs, the cost of sustainable operation and maintenance of water supply service, and the potential for reinvestment in the infrastructure. Moreover, to address the needs of the poor households the government either identify the poorest and subsidize their consumption or the community members pay for them.

Disagreement mostly arises between consumers who prefer to pay less and service providers who lean towards having a higher tariff level for stable revenue generation. The analysis of tariff collection rates in three WUAs for the last 5 years demonstrates that the average annual tariff collection rate stands at approximately 80%, which is considered satisfactory for the cost recovery and expansion. The highest tariff collection was recorded as 94% in Tezgari Poyon in WUA in Rudaki district in 2018 with the lowest being 71% in Delolo in Muminabad district in 2019 (Figure 9.4).

Social accountability. All WUAs have confirmed that developing and maintaining continuous dialogue with consumers and the district government are

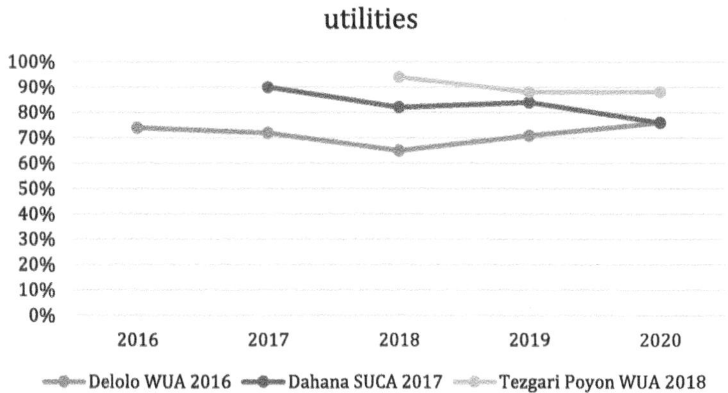

Figure 9.4 Tariff collection rate (*ibid.*).

critical to success. This has helped them raise more awareness about issues that they would like to solve and develop solutions in close coordination with community members. From 2016 Oxfam began to establish a social accountability mechanism within water utilities as part of the project funded by the Global Platform for Social Accountability of the World Bank Group. Social accountability can be defined as an approach towards building accountability that relies on civic engagement, that is, in which it is ordinary citizens and/or civil society organizations who participate directly or indirectly in executing accountability (Malena *et al.*, 2004). The project aimed at improving responsiveness and accountability in service delivery by supporting the service users to act collectively to influence key decisions, monitor service quality and demand better services.

Given that all households are shareholders in the water scheme, they understand that for the effective and efficient use of water resources, a joint decision-making body, in this case, the CABs, should act as the authorization platform to set preconditions for economically viable operations of WS systems. Oxfam has urged both sides to act as the owners and investors of the system and provide necessary financial support to minimize external dependence and ensure decentralized management. The purpose of the project was to provide a basis for constructive engagement between service users, service providers, and government institutions for sharing information on service performance, discussing discrepancies and issues, and identifying solutions that can be implemented through joint action.

Despite water utilities' inability to address the complaints adequately due to financial capacity, the assessment of data about the number of complaints received versus resolved shows very promising progress. As detailed in Figure 9.5, over the course of project implementation from 2016 to 2018, there

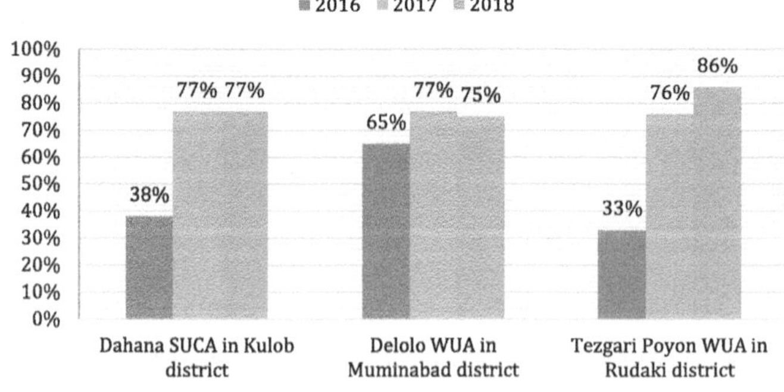

Figure 9.5 Customer satisfaction rate by rural water utility service.

was continuous improvement in customer satisfaction rate across three water utilities because of the application of a social accountability mechanism through CABs within each water utility.

9.8 CONCLUSION

By promoting sustainable business models and by improving the financial and managerial capacity of water utilities, there is a high chance that the WS service will be more resilient against disasters and be better prepared to respond when resources are dwindling. The programme analysis has shown that building such systems requires a blended funding mechanism and ownership by all interested parties, in a context where services had previously been provided for free and without a responsiveness to environmental risk. If the profit making is not secured in the service delivery, there is a slim chance of managing the WS system professionally on a voluntary basis. Moreover, the selected communities' economic capacity (demand) and water utilities' access to finance (supply) should also not be ignored.

In this chapter, a WASH business model and the vulnerability of the WS system to hazards in rural areas has been demonstrated. The key learnings from these activities are summarized as follows:

(1) *Creating demand:* It is highly important to assess the economic capacity of potential communities and their willingness to pay for water and sanitation services by monitoring or collecting data about the employment rate, key production areas, social activities and private/public health habits before their qualification for investment projects.

(2) *Developing supply:* It is equally critical to re-examine the profitability of water utilities in light of hazards and risks. Regular capacity-building activities and business know-how by sponsors or the government would be very valuable for WS service providers in integrating disaster risks into their budget planning and developing a contingency plan to act proactively against unforeseen cases. Quick access to finance marks a milestone of the capability of service providers to overcome challenges without delay thereby avoiding additional cost due to delay.

(3) *Enabling environment:* It is of utmost importance to create an institutional culture in which water utilities can perform based on realistic metrics that will incentivize profit making. Moreover, it is equally important to work with the government to align or adjust policies in line with the interests of those who will be willing to invest or manage the WASH services, especially in rural areas. In the future, opportunities for water utilities to access micro-loans from banks should be explored.

Future sustainability and resilience of WSS in rural Tajikistan depends on the following opportunities and risks:

Opportunities:

- **Sense of community ownership.** The community engagement and social accountability mechanism (CAB) anchors a great sense of ownership for the sustainability of the system.
- **Gradual tariff increase in line with inflation rate.** Increases in water price might generate revenue for water utilities further enabling them to address water issues and grow their customer base.
- **Decentralized service delivery.** The management model applied in rural areas does not require significant investments and is more cost-effective than connection to the centralized system.

Risks:

- **Local human resources to maintain the WS system.** All water utilities report high turnover and the loss of staff trained under the projects to operate and maintain the service delivery.
- **Governmental buy-in.** Government agencies have significant potential to impact project sustainability either positively or negatively. However, what is unclear is the government's financial contribution (or subsidy) in maintaining the WS system. The government buy-in in water service delivery in rural areas is essential to promote future expansion and quality application.
- **Aged infrastructure.** As consumers become more engaged in management processes and advocate for their rights, they become accustomed to getting stable and quality service from WUAs. However, as the infrastructure ages the re-investment responsibility lies outside the control of WUAs and users. The government has to step in and ensure that the state funding is within reach to rehabilitate the infrastructure in every 10–20 years to avoid any disruption in the service delivery and curb the public trust.

The current economic trend in Tajikistan requires a solid analysis of market size, prices, costs and returns for new actors to intervene in the WASH sector. The market challenges are still prevalent in involving potential investors in financing the WASH sector for profit. Besides, experiences in Tajikistan show that there is still highly limited knowledge regarding the principles of a market-based and decentralized WASH programme and the required design/implementation approaches. Moreover, most INGOs and donors have not yet shifted from supply-driven WASH programme delivery approaches in Tajikistan. The majority still provide funds for the implementation of WASH infrastructure without confirming that a payment system is in place to ensure the system's longer-term sustainability.

In conclusion, managing the WASH programme requires a combination of methodologies and analytical tools to assess the economic, social and environmental costs and benefits of WS infrastructure and service delivery at the rural level. Given that the WS service is a capital-intensive sector, achieving a resilient water supply in rural Tajikistan is especially challenging due to climate change, financial inflation and growing demographic trends. It requires investment not only in infrastructure, but also in institutional development of WUAs and data collection software for informed decision-making. The WUAs can deliver expected economic and health benefits to the community only when they are backed by appropriate support systems such as customer and government financing, skilled staff, an accountability mechanism and information. Addressing this requires long-term strategic district WASH planning with clear investment pathways that increase the government and communities' resilience to adapt over time in response to environmental risks and developments.

REFERENCES

Assembly UG (2016). Report of the open-ended intergovernmental expert working group on indicators and terminology relating to disaster risk reduction, United Nations General Assembly, New York, NY, USA.

Law of the Republic of Tajikistan N1633 (2019). Закон Республики Таджикистан от 19 июля 2019 года № 1633 «О питьевом водоснабжении и водоотведении» (Law of the Republic of Tajikistan N1633 on 'Drinking Water and Wastewater', 19 July 2019).

Malena C., Reiner F. and Janmejay S. (2004). Social Accountability: An Introduction to The Concept and Emerging Practice, Social Development Paper 76. World Bank, Washington, DC.

Matveeva A. (2009). Working Paper no. 46, The perils of emerging statehood: civil war and state reconstruction in Tajikistan, An Analytical Narrative on State-Making, Crisis States Research Centre, March 2009, p. 4. Available at https://assets.publishing. service.gov.uk/media/57a08b7bed915d3cfd000d4e/wp46.2.pdf (accessed 22 January 2021).

Mustaeva N., Wyes H., Mohr B. and Kayumov A. (2015). Tajikistan: Country Situation Assessment. Working Paper, CAREC, p. 30. Available at https://idl-bnc-idrc. dspacedirect.org/handle/10625/57346 (accessed 27 January 2021).

Oxfam Management Information System (MIS), last updated in January (2021).

Oxfam Quarterly MEAL and Technical Review Report (2021).

Resolution # 679 of the Government of Republic of Tajikistan (2011). Постановление Правительства Республики Таджикистан от 31 декабря 2011 года №679 «Об утверждении Порядка государственного контроля и надзора питьевого водоснабжения», http://www.cawater-info.net/pdf/tj-679-2011.pdf, (Resolution # 679 of the Government of Republic of Tajikistan dated December 31, 2011).

Tajikistan-World Bank Group Country Partnership Framework 2019–2023. (15 May 2019). Available at https://www.worldbank.org/en/country/tajikistan/publication/cpf-2019-2023 (accessed 27 January 2021).

UN Water (2021). SDG6 Snapshot in Tajikistan, Available at https://sdg6data.org/country-or-area/Tajikistan (accessed 27 January 2021).

UN-Water, SDG 6. Available at https://www.sdg6data.org/. (accessed 30 January 2021).

World Bank Group (2017). Glass Half Full: Poverty Diagnostic of WS, Sanitation, and Hygiene Conditions in Tajikistan. World Bank, Washington, DC, p. 1. © World Bank. Available at https://openknowledge.worldbank.org/handle/10986/27830 (accessed 22 January 2021).

Chapter 10
Conclusions

Leslie Morris-Iveson and St John Day

This book has provided a series of reflections by practitioners on how to make advancements into building resilient water supply services, on the basis of practical application. The main chapters include cases from a range of different contexts. This includes localities where consumers are broadly happy with their water supply and providers are planning ahead to address growing pressures. Real examples are also provided from areas where satisfaction is much lower, where substantial parts of the population do not have safely managed access to services at all. Some of the programmes and initiatives have been developed with high visibility, and with sufficient resources (including funding); however, some have received very little input, in terms of support and capacity. In all examples, action needs to be taken to improve performance levels, systems need to be in place to respond rapidly to threats and longer-term planning must ensure water security for customers and consumers; as well as safeguarding the natural environment.

The case studies do not set out to portray what resilience looks like in practice – but they are representative of the ongoing efforts to respond better to the threats represented by practitioners themselves. We hope the cases can stimulate a realistic and broad discussion on actions currently taking place more generally – to understand how to respond better in the future. They provide points to ponder on how to further develop their own innovative approaches and solutions.

Resilience might be a concept that many understand and recognise the need to pursue, but on the ground deciding what actions to take can be challenging. The cases have shown a diversity of solutions that contribute to resilience, and that in some way ensure that water supply continues to reach people despite risks. In reality, service providers will need to pursue and balance a range of actions across keeping water demands in check. Sustainably augmenting water supply, for example, will always need to go hand in hand with demand management practices. Collaborating with diverse stakeholders across the sector and with other uses of water will need to take place as well. A coherent approach that encompasses not only a diversity in technologies, but also in the types of approaches will be needed that links corporate and financial elements (not discussed in this book), adding another layer of complexity.

A final lesson has shown us that as many of the cases were described prior to the coronavirus disease 2019 pandemic, changes in the context have given even further testing of resilience approaches to ensure water keeps flowing during the pandemic – a necessity to ensure people continue to benefit from water and to practice good hygiene (i.e. handwashing), one of the measures cited to stem transmission. The contributors to this book have reported delays in actions implemented, particularly in capacity building due to movement restrictions, and difficulty in working with stakeholders in many contexts where face-to-face communication is essential.

Drawing from a range of contexts and threats, as editors of this book our main recommendations focus on the following six aspects.

- Water companies and service providers need to ensure they have a vision and a sound grasp of local, national and global threats that undermine all aspects of water supply performance. This needs to result in appropriate and effective interventions, including implementing the best selection of technologies, approaches, and partnerships, and focussing on a few interventions with the biggest impacts. One approach will not hold all the answers to resilience, a broad range of responses are needed. We refer to this as taking the correct action, at the right time, and doing it well.
- To function effectively, front-line operators continuously need to develop and fine-tune approaches to respond to chronic, rapid and slow-onset threats. For example, water supply, sewage and other essential infrastructure is often not spared the impacts of rapid disasters. All aspects of water treatment and supply can be affected and this has a direct impact on households, neighbourhoods, schools, hospitals and other essential services. Front-line operators need to be able to respond rapidly to fix, repair or even just patch-up critical assets, as well as looking at future threats like climate change and growing demand.
- Often practitioners work under incredibly difficult and testing circumstances. There needs to be a focus on *doing what you can with what you have*. When major water supply infrastructure is already absent or inadequate, emergency

interventions may only provide a stop-gap until existing assets are rehabilitated or new physical assets are provided. Asset management planning to build resilience and strengthen utility performance needs to be better documented. This will enable suppliers to adopt a systematic approach to meet the required service levels in the most cost-effective manner. It will also help water companies to justify interventions to potential investors.

- Front-line operators need greater support to document and publish their detailed knowledge and experience, which they have in abundance. This book has tried to demonstrate this benefit, but more data, better information and real-time customer feedback is required.
- Greater support for collaboration – across boundaries and between sectors – is necessary for structured cross-learning. Understanding different experiences can lead to new insights, leading actors to enable new capacities and initiatives. This book might be a first step in seeing what actions emerge, and what they look like in very different contexts, building on this would be to build capacities and skills that enable new initiatives. As we strive to make water supplies more effective and resilient there needs to be greater transparency, openness and commitment to ongoing adaptation and learning.
- Although many different threats have been described, water scarcity has been specifically highlighted in all cases as a major threat to sustainable water supply. Globally, a number of responses are in place. However, the conclusions drawn by front-line suppliers point to the urgent necessity for longer term planning frameworks, and sufficient resources and capacities to address this universal threat. System-wide efficiencies are needed to address this threat.

Although there is no 'one size fits all' solution, what is clear is that in order to build more resilient services, interventions need to be both *appropriate* and *effective*–delivered to high professional standards, and requiring various stakeholders to work together. In contexts where high service levels already exist then practitioners can often take action on specific and targeted interventions to address current and future pressures. However, if water companies are starting from a low base, or the network of people, institutions, infrastructure, policy and finance are absent, then they need to focus on interventions that will achieve the greatest impacts. Often these water companies face a range of challenges, with perhaps the most obvious being catchment degradation, inadequate infrastructure, poor service performance levels, and limited revenue generation and investment. All of these inter-connected problems need to be addressed and they will not be overcome by focussing solely on climate resilience.

Resilience is an evolving concept, and approaches on the ground are constantly developing. As capacities grow and water suppliers can anticipate and mitigate risks to water supply we will be better prepared to deal with the challenges that water supply faces today.

Index

IWA Publishing's authorised EU representative for General Product
Safety Regulations is Diane D'Arras, 15 rue Duret, 75116 Paris,
France, e-mail: safety@iwap.co.uk.

Printed and bound by CPI Group (UK) Ltd, Croydon, CR0 4YY
12/05/2026
02108870-0002